AF364250

Medicinal Chemistry

Drugs Acting on Nervous System

Medicinal Chemistry
Drugs Acting on Nervous System

Kapil Kalra

Deepak Nanda

Dev Bhoomi Institute of Pharmacy & Research Dehradun,
Uttarakhand

PharmaMed Press
An imprint of Pharma Book Syndicate

An unit of BSP Books Pvt., Ltd.

4-4-316, Giriraj Lane,
Sultan Bazar, Hyderabad - 500 095.

Published by

PharmaMed Press
An imprint of Pharma Book Syndicate

An unit of BSP Books Pvt., Ltd.

4-4-316, Giriraj Lane, Sultan Bazar, Hyderabad - 500 095.
Phone: 040-23445605, 23445688; Fax: 91+40-23445611
E-mail: info@pharmamedpress.com
www.pharmamedpress.com/pharmamedpress.net

ISBN : 978-93-89974-82-9

Preface

THIS BOOK is meant for undergraduate students of pharmacy. We have tried to incorporate the latest developments in the book.

We are grateful to our Colleagues Mr. Pankaj Nainwal, Mr. Amandeep, Mr. Ravindra P.Singh, Ms. Manisha Gaur and Mr.Yogesh for their moral and material support throughout the preparation of the manuscript of this book. Without their inspiration and support, it would not have been possible for us to complete this task.

Special Mention needs to be made for constant encouragement and experts guidance exerted by Mr. Sanjay Bansal (Chairman DBGI Dehradun), Prof. (Dr.) B.K. Razdan, Dr. T.S.Parmar, Prof. (Dr.) N.V Satheesh Madhav, Prof.(Dr.) Vijay Juyal, Prof.(Dr.) A.K.Sharma, Prof.(Dr.) Firoz Anwar, Prof.(Dr.) Kumud Upadhaya.

- Authors

Contents

UNIT – I

AUTONOMOUS NERVOUS SYSTEM

- Cholinergic, Anticholinergic and Anticholinesterases
- Adrenergic Drugs and Antiadrenergic Drugs
- Neuromuscular Blocking Agents
- Drugs used in the Treatment of Alziemer's Disease
- Local Anaesthetics

AUTONOMOUS NERVOUS SYSTEM

Cholinergic Drugs and Related Agents

The autonomic nervous system (ANS) is composed of two divisions-sympathetic and parasympathetic. Acetyl choline serves as a neuro transmitter at both sympathetic and parasympathetic pre-ganglionic nerve endings.

Cholinergic agents are drugs that either directly or indirectly produce effect similar to acetyl choline (Ach). The neurotransmitter of pre ganglionic neuron is acetyl choline and post ganglionic neuron is nor adrenaline in sympathetic system. Acetyl choline is the neuro transmitter of all pre and post ganglionic neurons of parasympathetic system.

Cholinergic Receptors

There are two types of cholinergic receptors on the basis of their ability to be bound by the naturally occurring alkaloids-nicotine and muscarine are called nicotinic receptors and muscarinic receptor.

1. Nicotinic receptors

Nicotinic receptors are coupled directly to ion channels and mediate very rapid responses when activated by acetyl choline. These receptors are selectively activated by nicotine and blocked by tubocurarine or hexamethonium. These belongs to ligand -gate ion channel receptors and acetyl choline serve as a gate keeper by interacting with the receptor to modulate passage of ions, principally K^+ and Na^+ through the channel. Their activation causes opening of the channel and rapid flow of cation resulting depolarization and generation of action potential.

Sub types: There are two types

N_1 *nicotinic receptors:* These are present in neuromuscular junction. They are blocked by succinyl choline, d – tubocurarine and decamethonium and stimulated by phenyl trimethyl ammonium.

N_2 *nicotinic receptors:* These are found in autonomic ganglia. They are blocked by hexamethonium and trimethaphan but stimulated by tetra methyl ammonium and diethyl 4-phenyl piperazinium (DMPP).

2. Muscarinic receptors

Muscarinic receptors play as essential role in regulating the functions of organs in ANS to maintain homeostasis of the organism. The action of acetyl choline on Muscarinic receptor can result in stimulation or inhibition of the organ system affected. Acetyl choline stimulates secretions from salivary and sweat glands, secretions contraction of the gut and constriction of the airways of the respiratory tract.

Sub types: There are five types of receptors M_1, M_2, M_3, M_4 and M_5.

1. M_1 *receptors:* They are located in CNS, exocrine glands and autonomic ganglia. These are identified in sub mucosal glands and some smooth muscles. When stimulated M1 receptors cause gastric secretion.

2. M_2 *receptors:* These are called cardiac muscarinic receptors because they are located in atria and conducting tissue of the heart. Their stimulation causes a decrease in the strength and rate of cardiac muscle contraction. M2 receptor activate K^+ channels to cause hyper polarization of cardiac cells, resulting in bradycardia.

3. M_3 *receptors:* These are referred to as "glandular" muscarinic receptor, are located in exocrine glands and smooth muscles. Glandular secretions from lacrimal, salivary, bronchial, pancreatic and mucosal cells in GI tract are characteristics of M_3 receptor stimulations.

4. M_4 *receptors:* They are present in tracheal smooth muscle, when stimulated inhibit the release of acetyl choline.

Bio Chemical Effects of Muscarinic Receptor Stimulation

Transmission of synapse involving second messenger is much slower compared with at synapses where ion channels are activated directly. The sequence of bio chemical events in this second messenger system begins with activation of the receptors by an agonist and involves the activation of G–proteins that are bound to a portion of the intracellular domain of the Muscarinic receptors.

G–Proteins consists of 3 sub units, α, β and γ. When the receptor is occupied, the sub unit which has enzymatic activity, catalyses the conversion of GTP to GDP (guanosine diphosphate) which can associate with various enzymes and ion channels. A single drug–receptor complex can active several G–protein molecules and each can remain associated with a target molecule.

Adenylate cyclase is a membrane enzyme is another target of muscarinic receptor activation. The second messenger cAMP is synthesized within the cell from adenosine tri phosphate (ATP) by the actions of Adenylate cyclase. The cAMP activate protein kinase which catalyze the phosphorylation of enzymes and ion channels, alter the amount of calcium entering the cell and thus affecting muscle contraction.

But muscarinic receptor activation causes lower levels of cAMP, reducing CAMP protein dependent kinase activating and relaxation of muscle contraction.

Stereo Chemistry of Acetyl Choline

Acetyl choline exists in number of confirmations. Confirmational isomers of acetyl choline derived from rotation around – O –C – C –N – axis. Four of these confirmations are illustrated by Newman projection below.

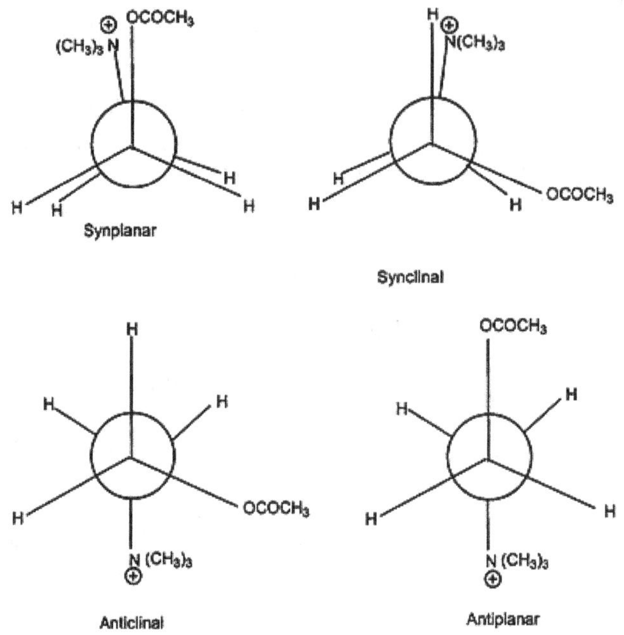

Cholinomimetic (or) Parasympathomimetic Drugs

These are agents that mimic the action at parasympathetic system.

Classification

1. **Directly acting Cholinergic drugs (Agonist)**

   A. *Choline Esters* – Acetyl choline, Carbochol, Bethanechol, Methacholine.

   B. *Alkaloids* – Pilocarpine

2. **Indirectly acting cholinergic drugs (Anti choline esterase)**

   A. *Reversible Inhibitors* – Physostigmine, Neostigmine, Pyridostigmine, Endrophonium chloride, Ambinonium chloride.

   B. *Irreversible Inhibitors* – Pralidoxime chloride, Isoflurphate, Echothiophate iodide, Parathion, Malathion.

Acetylcholinesterase Inhibitor (Anticholine esterases)

An **acetylcholine esterase inhibitor** (often abbreviated AChEI) or **anti-cholinesterase** is a chemical that inhibits the choline esterase enzyme from breaking down acetylcholine, increasing both the level and duration of action of the neurotransmitter acetylcholine.

Uses

- Occur naturally as venoms and poisons

- Are used as weapons in the form of nerve agents

- Are used medicinally.

  - To treat myasthenia gravis. In myasthenia gravis, they are used to increase neuromuscular transmission.

  - To treat Alzheimer's disease

  - To treat Lewy Body Dementia

  - As an antidote to anticholinergic poisoning

A. Acetylcholine Chloride

$$H_3C - \overset{\overset{\displaystyle CH_3}{|}}{\underset{\underset{\displaystyle CH_3}{|}}{\overset{+}{N}}} - CH_2 - CH_2 - O - C - CH_3 \quad \overset{-}{Cl}$$

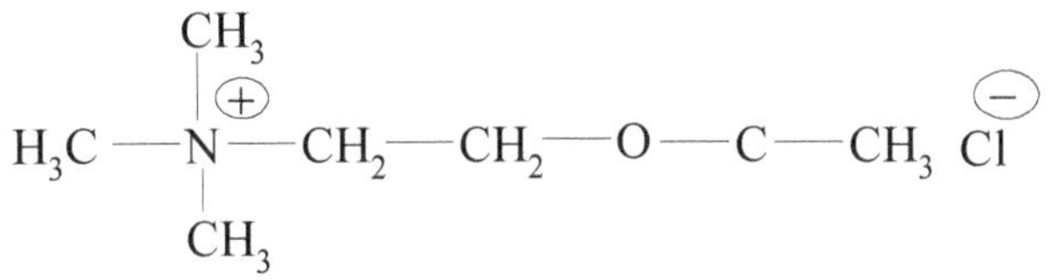

(2- Acetyloxy ethyl) - trimethyl ammonium chloride

Synthesis

$$(CH_3)_3 N + CH_3 COOCH_2CH_2 Cl \longrightarrow (CH_3)_3 - \overset{+}{N} - CH_2 CH_2 - COOCH_3 \quad \overset{-}{Cl}$$

Trimethyl amine   2 - Chloro ethyl acetate   Acetylcholine chloride

Neostigmine Bromide

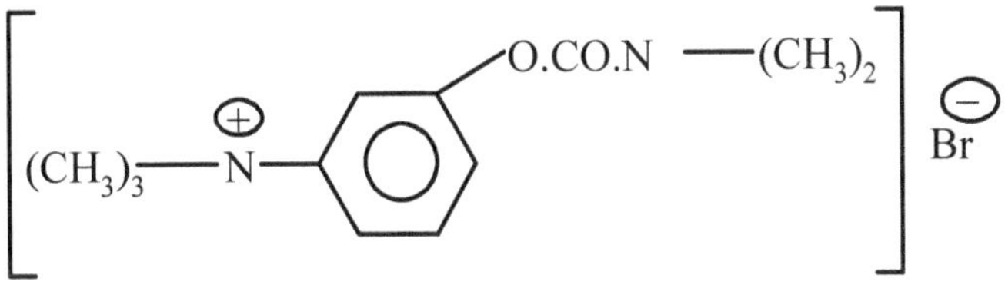

$$\left[(CH_3)_3 - \overset{+}{N} - \underset{}{\bigcirc} - O.CO.N - (CH_3)_2 \right] \quad \overset{-}{Br}$$

3 -{ [(Dimethylamino) carbonyl] oxy} - N, N, N - trimethyl benzeneammonium bromide

Synthesis of Neostigmine

Neostigmine was first synthesized by Aeschlimann and Reinert in 1931. Neostigmine is made by first reacting 3-dimethylaminophenol with N-dimethylcarbamoyl chloride, which forms a dimethylcarbamate. Next, that product is alkylated using dimethylsulfate, which forms neostigmine.

Pilocarpine

Pilocarpine is a muscarinic alkaloid obtained from the leaves of tropical American shrubs from the genus *Pilocarpus*. It is a non-selective muscarinic receptor agonist in the parasympathetic nervous system, which acts therapeutically at the muscarinic acetylcholine receptor M_3 due to its topical application, e.g., in glaucoma and xerostomia.

(3S-cis) - 3- Ethyl dihydro - 4-[1-methyl-1 H-imidazol-Syl) methyl] 2(3H) furanone

Clinical Uses

Pilocarpine has been used in the treatment of chronic open-angle glaucoma and acute angle-closure glaucoma for over 100 years. It acts on a subtype of muscarinic receptor (M_3) found on the iris sphincter muscle, causing the muscle to contract and produce miosis. Pilocarpine also acts on the ciliary muscle and causes it to contract. When the ciliary muscle contracts, it opens the trabecular meshwork through increased tension on the scleral spur. This action facilitates the rate that aqueous humor leaves the eye to decrease intraocular pressure. Pilocarpine is often used as an antidote for Scopolamine, Atropine and Hyoscyamine poisoning.

Methacholine

Methacholine chloride (trade name **Provocholine**) is a synthetic choline ester that acts as a non-selective muscarinic receptor agonist in the parasympathetic nervous system.

The primary clinical use of methacholine is to diagnose bronchial hyperreactivity, which is the hallmark of asthma and also occurs in chronic obstructive pulmonary disease. This is accomplished through the bronchial challenge test. Other therapeutic uses are limited by its adverse cardiovascular effects, such as bradycardia and hypotension, which arise from its function as a cholinomimetic.

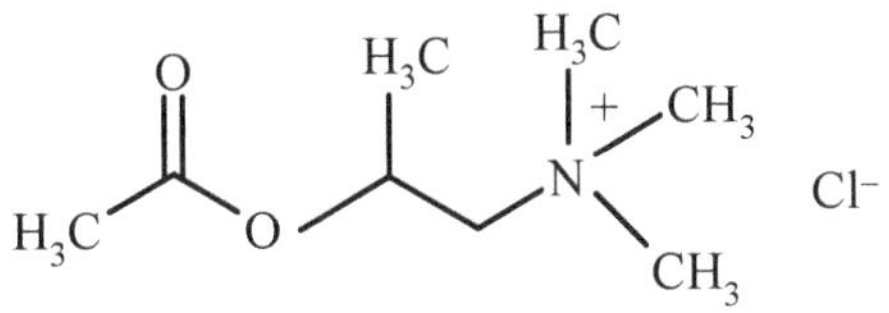

Physostigmine

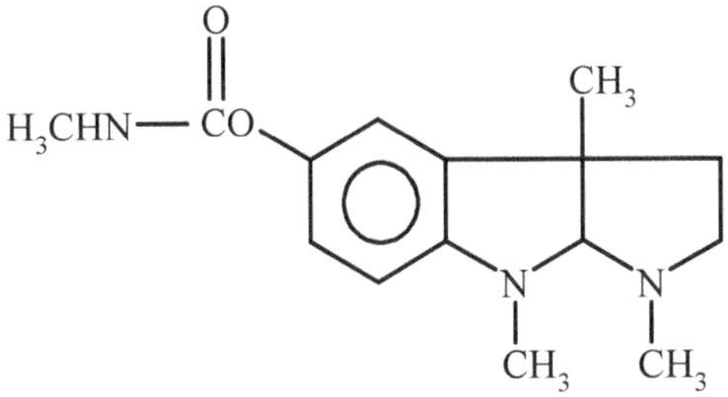

Pyrrolo [2, 3 - 6] indol - 5- ol - 1, 2, 3, 3a, 8, 8a
- hexahydro - 1, 3a 8 - trimethyl methyl carbamate

Synthesis of Physostigmine

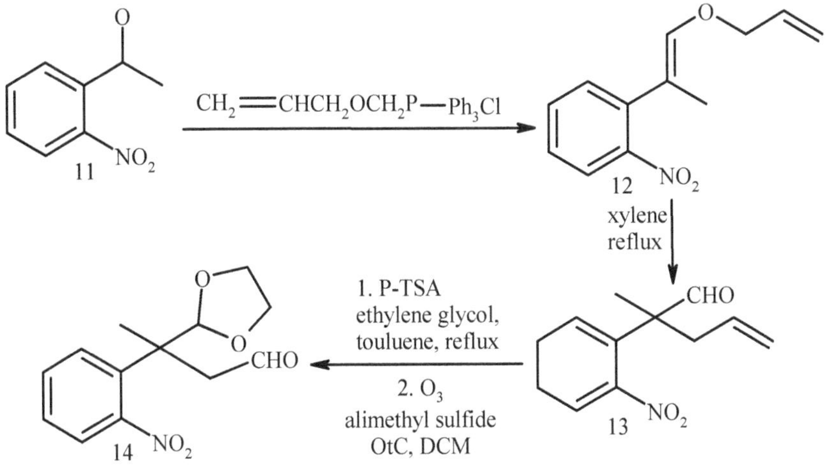

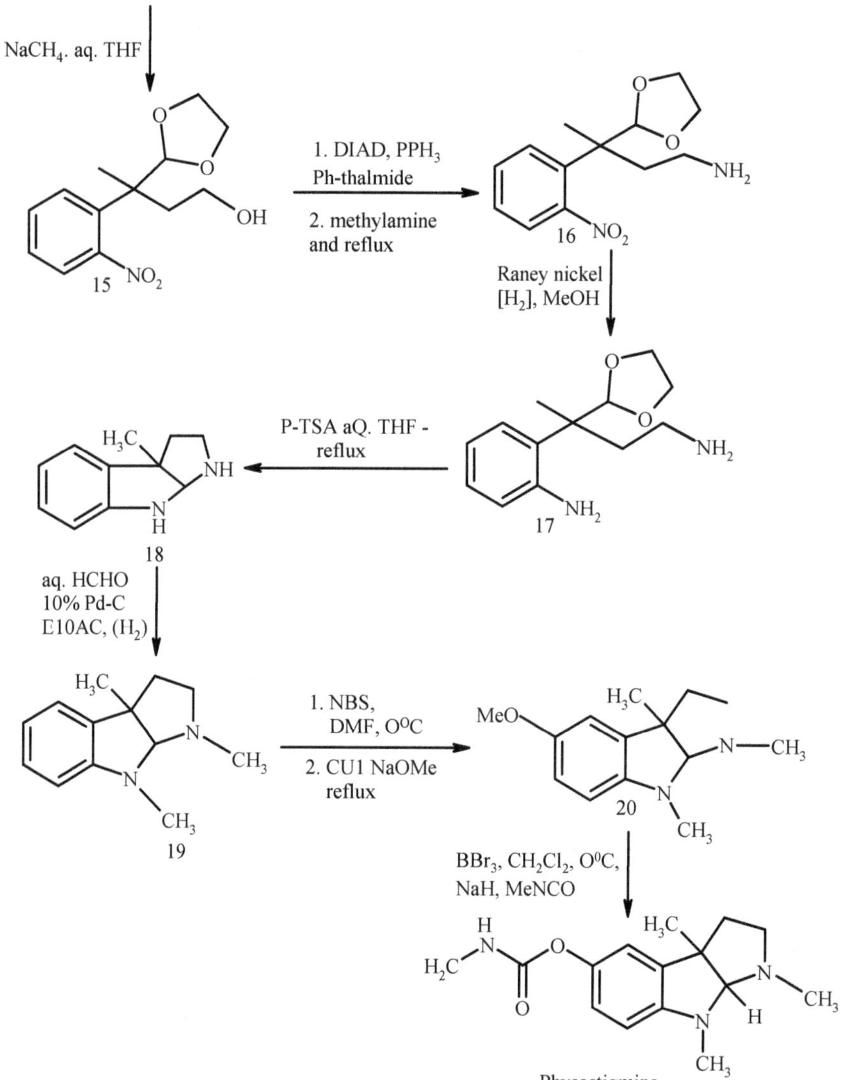

NaCH4. aq. THF
O
O
OH
NO2
15
1. DIAD, PPH3
Ph-thalmide
2. methylamine
and reflux
O
O
NH2
16 NO2
Raney nickel
[H2], MeOH
O
O
NH2
NH2
17
P-TSA aQ. THF -
reflux
H3C
NH
N
H
18
aq. HCHO
10% Pd-C
E10AC, (H2)
H3C
N
CH3
N
CH3
19
1. NBS,
DMF, O0C
2. CU1 NaOMe
reflux
MeO
H3C
N—CH3
N
20
CH3
BBr3, CH2Cl2, O0C,
NaH, MeNCO
H
N
H2C
O
O
H3C
N
CH3
N
H
CH3

Physostigmine

SAR for Cholinergic Drugs

SAR depends upon the modifications of the following groups

N - CH2 - CH2 - O - C - R
O
Ester function
Choline
Onuim

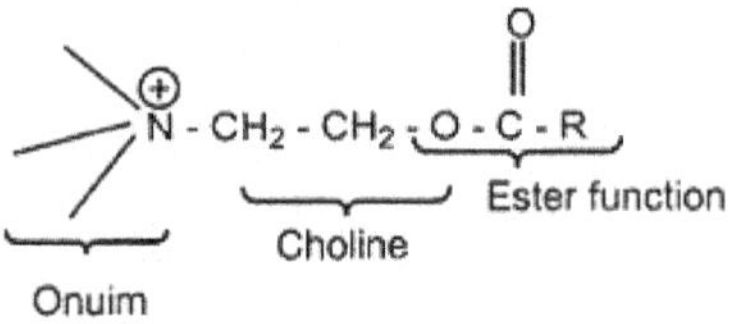

2- [(Amino carbanoyl) oxy] - N,N,N - trimethyl ethanamium chloride

I. Quaternary ammonium group

1. The onium group is essential for intrinsic activity and contributes to the affinity to receptors.

2. Replacement of nitrogen with sulfur, arsenic or selenium produces less active compounds.

3. Primary, secondary or tertiary amines are less active than acetyl choline.

4. Replacement of methyl groups by ethyl or larger alkyl groups produces inactive compounds.

5. Compounds in which two methyl groups on nitrogen were retained and they were replaced by a larger alkyl groups were found to have considerable activity.

II. Modification of Ethylene Bridge

1. Shortening or lengthening of the ethylene group that separates the ester group and ammonium group reduces muscarinic activity.

2. An α- substitution decreases nicotinic activity more extent.

3. Replacement of hydrogen atoms of ethylene bridge by alkyl groups are less active except when a single methyl group is placed either at α or β to the quaternary nitrogen atom.

4. The presence of methyl group α to nitrogen increases nicotinic activity (Acetyl methyl choline) and β to nitrogen increases muscarinic activity (Methacholine)

5. Hydrolysis by acetyl cholinesterase is more affected by β–substitutions than α–carbon.

III. Modification of ester group

1. The ester group of acetyl choline contributes to the binding of the compound to the muscarinic receptor.

2. When acetyl group is substituted by its higher homologues produce less active compounds.

3. When the acetyl group is replaced by a carbonyl (carbachol) has both muscarinic and nicotinic property.

4. Esters of aromatic or higher molecular weight acids possess cholinergic antagonist activity.

5. The methyl ester is rapidly hydrolyzed by cholinesterase.

6. When the terminal methyl group is replaced by NH_2 group, the resulting compound is potent cholinergic agent with both muscarinic and nicotinic activities.

Anticholinergics

Cholinergic Blocking Agents, Cholinergic Receptor Antagonists (or) Anti Spasmodics

The anti cholinergic drugs are agents that inhibit the effect on Acetyl choline (Ach) released from post ganglionic parasympathetic nerve endings. They block muscarinic action of Ach including smooth muscle contraction and exocrine gland secretion. Because of the ability to relax smooth muscles, they are also referred as antispasmodics. The cholinergic blocking agents that have high affinity to the receptors may decrease the no. of available free receptors and the efficiency of the endogenous neuro transmitter like Ach. Anti muscarinic drugs act by competitive antagonism of Ach binding to muscarinic receptors.

Classification

1. **Solanaceous alkaloids and Analogues:** Atropine Sulfate, Hyoscyamine sulfate, Scopalamine HBr, Homatropine HBr, Ipratropium bromide.

2. **Amino alcohol esters:** Cyclopentolate HCl, Clidinium bromide, Dicyclomine HCl, Glycopyrrolate, Methanthelin bromide, Propanthelin bromide, Mepenzolate.

3. **Amino Alcohols:** Biperidine HCl, Procyclidine HCl

4. **Amino alcohol ethers:** Benztropine mesylate, Orphenadrine

5. **Amino amide:** Tropicamide, Isopropamide iodide

6. **Miscellaneous:** Ethopropazine.

Atropine

Benzene endo ($\pm$) -α- (hydroxy methyl) -8-methyl azabicyclo [3-2-1] oct - 3 - yl acetate

Synthesis

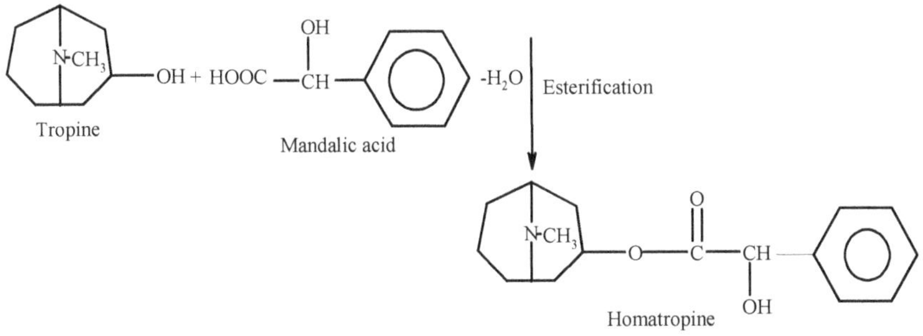

Cyclopentolate Hydrochloride

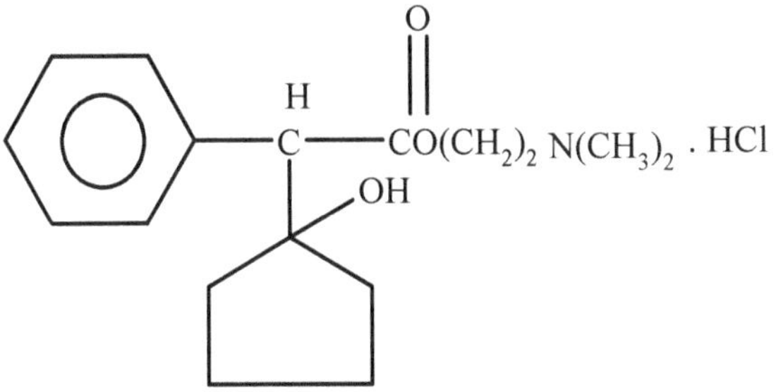

2- (Dimethylamino) ethyl-1-hydroxyl-α-phenyl
cyclopentane acetate hydrochloride

SAR for Cholinergic Blocking Agents

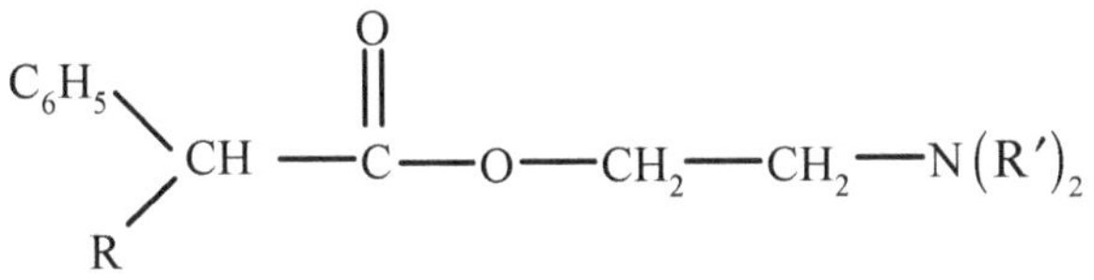

1. Anti cholinergic compounds has some structural similarity to acetyl choline but contain additional substituents which enhance their binding to cholinergic receptors.

2. R- may be hydroxy alkyl, alkyl, cycloalkyl or heterocyclic group for good anti cholinergic activity.

3. The nitrogen is tertiary atom which contains alkyl group not larger than butyl for effective antagonist activity.

4. The acyl group is always larger than acyl group in acetyl choline for good activity. Hydrophobic substituents increase the affinity to binding the receptors and have good antagonist property.

5. The presence of free hydroxyl or carbamide is also important for hydrogen bonding with receptor.

6. Naturally occurring l-hyocyamine is more active than d-isomer.

Mechanism of Action

The cholinergic blocking agents are competitively inhibit the cholinergic receptors and prevent the binding of acetylcholine to the receptors due to the size of acyl group through 'umbrella effect'.

The large group (alkyl or aryl) present in cholinergic blocking agents increase the affinity of the blocking agent and also block the approach of acetylcholine to the receptor.

Adrenergic Drugs

Adrenergic Receptors

Sympathomimetic or Adrenergic drugs exert their effects by direct action on adrenergic receptors. There are two types.

1. alpha-Adrenergic receptors

2. beta- Adrenergic receptors

1. Alpha-Adrenergic Receptors

Alpha-Adrenergic receptor sites have three parts

(i) Anionic site (Phosphate) which binds with +ve ammonium group.

(ii) One hydrogen binding area

(iii) One flat area for aromatic ring binding.

Nor epinephrine activates primarily alpha-Adrenergic receptors

Alpha-Adrenergic receptors have two types

(a) *alpha-1-Adrenergic receptors-* which are found in smooth muscles of iris, arteries, arterioles and veins. It exerts their effect on post synoptic nerves. Alpha1-Adrenergic receptor activation increases the influx of extracellular Ca^{2+} at calcium channels.

(b) *alpha-2-Adrenergic receptors* – which mediate the inhibition of adrenergic neuro transmitter release. It exerts their effect on pre synoptic nerves. Activation of alpha-2-Adrenergic receptors leads to a reduction in the catalytic activity of adenylyl cyclase.

2. **Beta- Adrenergic receptors**

Beta- Adrenergic receptor sites have the following parts

(i) Anionic site which binds with +ve ammonium group.

(ii) One hydrogen binding area

(iii) One flat area for aromatic ring binding.

Epinephrine activates primarily beta- Adrenergic receptor. Beta- receptor activation relaxes bronchial smooth muscles which cause bronchi of the lungs to dilate and also increases the rate and force of heart contractions.

Beta- Adrenergic receptors are three types. They are

(a) *beta-1-Adrenergic receptors*: found in myocardium where their stimulation increases the rate and force of myocardial contractions. They are located mainly in the heart, where they mediate the +ve inotropic and chronotropic effects of the catecholamines. They exhibit the agonist potency in the order of Isopreterenol > Epinephrine = Nor Epinephrine.

(b) *beta-2-Adrenergic receptors*: found in bronchial and vascular smooth muscles where their stimulation causes smooth muscle dilatation and relaxation. They exhibit the agonist potency in the order of Isopreterenol > Epinephrine > Nor Epinephrine.

(c) *beta-3-Adrenergic receptors*: They are expressed on fat cells and their stimulation causes lipolysis. They are located on brown adipose tissue and is involved in the stimulation of lipolysis. They exhibit the agonist potency in the order of Isopreterenol = Nor Epinephrine > Epinephrine.

Classification of Adrenergic Drugs

1. *Direct acting Adrenergic Agonists*: They are bind and activate a1, a2, b1, b2 receptors

Examples: Nor Epinephrine (binds with a1, a2, b1, receptors), Epinephrine (binds with a1, a2, b1, b2, receptors), Dopamine (with a1, a2, b1, b2, receptors), Xylometazoline, Phenyl ephrine, Methoxamine, Isoprenaline, Salbutamol.

2. *Indirect acting Adrenergic Agonists*: They produce Nor Ephidrine (NE) like actions by stimulating NE release, preventing its reuptake and thus its inactivation.

Example: Tyramine

3. *Dual acting Adrenergic Agonists*: These agents act as direct and indirect adrenergic agonists. They bind to adrenergic receptors and stimulate NE release.

Example: Ephedrine, Amphetamine, Mephentermine.

Chemical Classification

1. *Phenyl ethylamines and related compounds*

   Adrenaline, Nor Adrenaline(NE), Dopamine, Phenyl ephrine, Methoxamine, Methyl dopa, Isoproterinol(Isoprenaline), Salbutamol, Terbutaline, Dobutamine, Amphetamine, Ephedrine, Pseudo Ephedrine, Ritodrine, Salmeterol

2. *Imidazoline derivatives*

   Naphazoline, Tetrahydrazoline, Oxymetazoline, Xylometazoline, Clonidine.

1. **Direct acting Adrenergic Agonists**

   A. Endogenous Catecholamines

   Epinephrine: (Adrenaline)

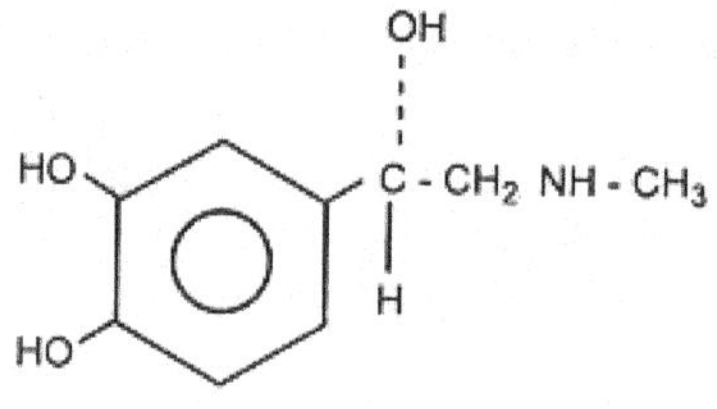

1 – (3, 4 - Dihydroxyphenyl) – 2- methyl amino ethanol.

Norepinephrine (Noradrenaline)

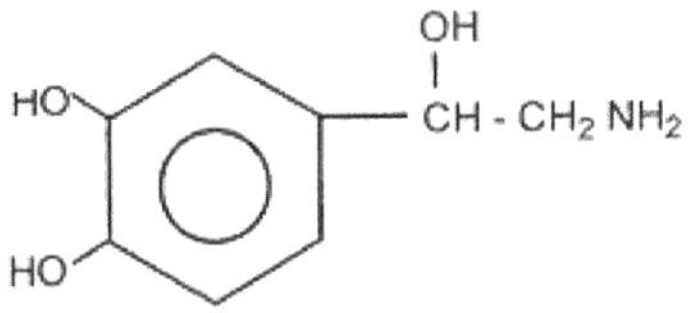

L - 1- (3, 4 Dihydroxy phenyl) - 2 - amino ethanol

Synthesis of Epindephrine

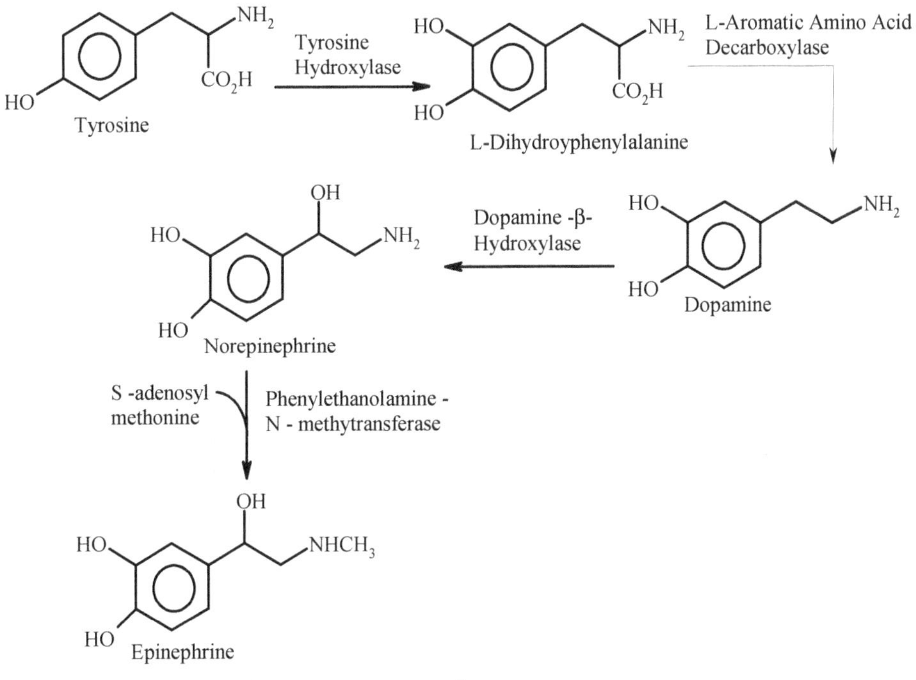

B. Alpha-Adrenergic Receptor Agonist

Phenylephrine

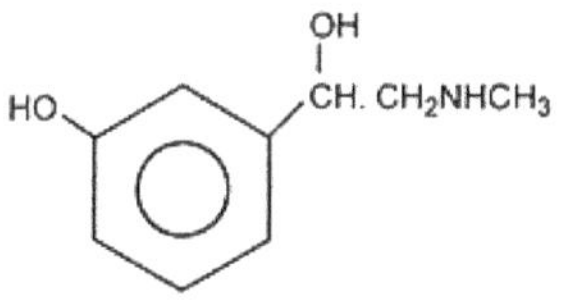

(-) - 1 - (3 – Hydroxyphenyl) – 2 – methyl amino ethanol

Synthesis

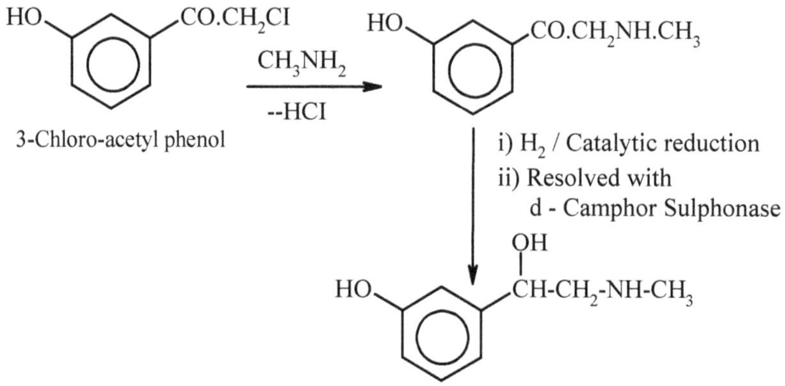
HO
CO.CH2CI
3-Chloro-acetyl phenol
CH3NH2
--HCI
HO
CO.CH2NH.CH3
i) H2 / Catalytic reduction
ii) Resolved with
d - Camphor Sulphonase
OH
HO
CH-CH2-NH-CH3

(–) Phenylphrine

Methyl Dopa

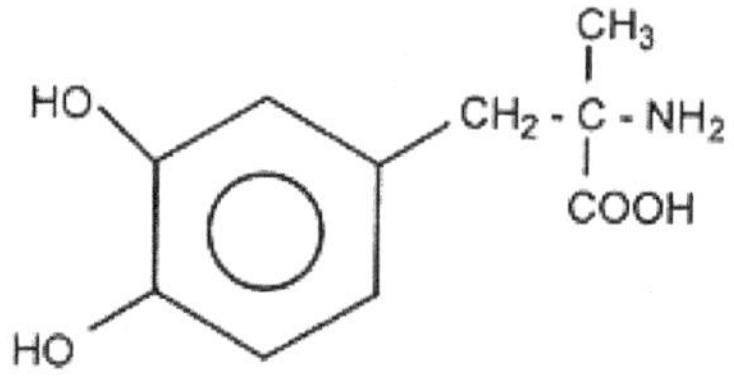
CH3
HO
CH2 - C - NH2
COOH
HO

L – α – methyl – 3, 4 - dihydroxy phenyl alanine.

Imidazolines

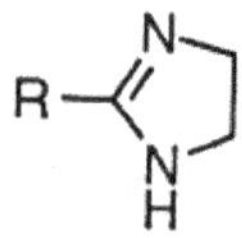
N
R
N
H

Naphazoline: R = 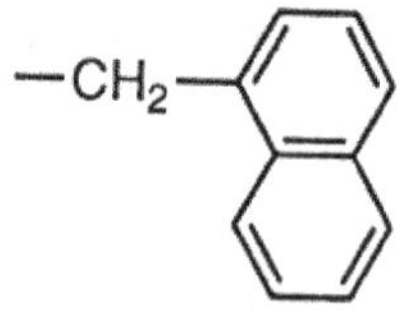
—CH2

Tetrahydrozoline: R = 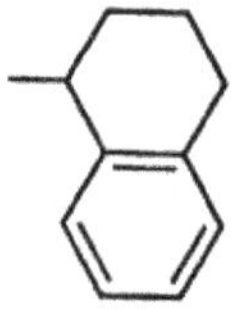

Isoproterenol (Isoprenaline)

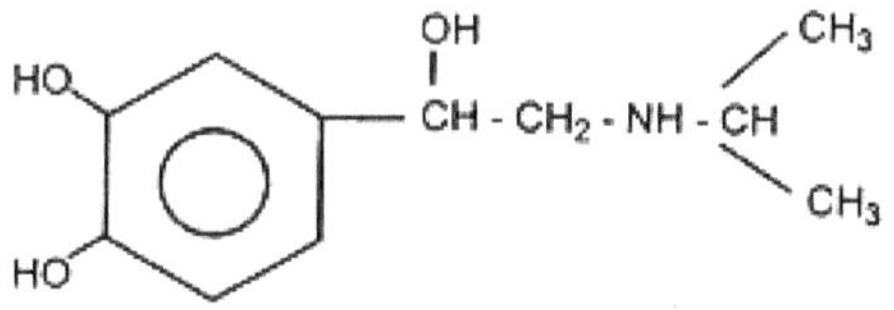

(dl) – β – (3,4 - Dihydroxy phenyl) – α – isopropyl amino ethanol.

C.   b-Adrenergic Receptor Agonist

Terbutaline

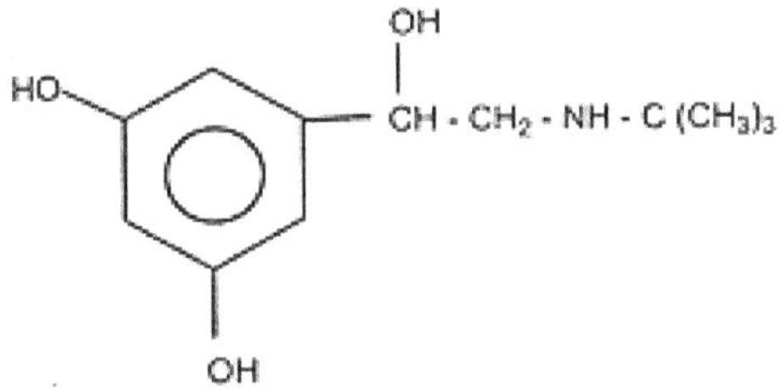

N – tert Butyl – 2 – [(3,5 - dihydroxy phenyl) – 2 – hydroxy] ethyl amine.

Synthesis

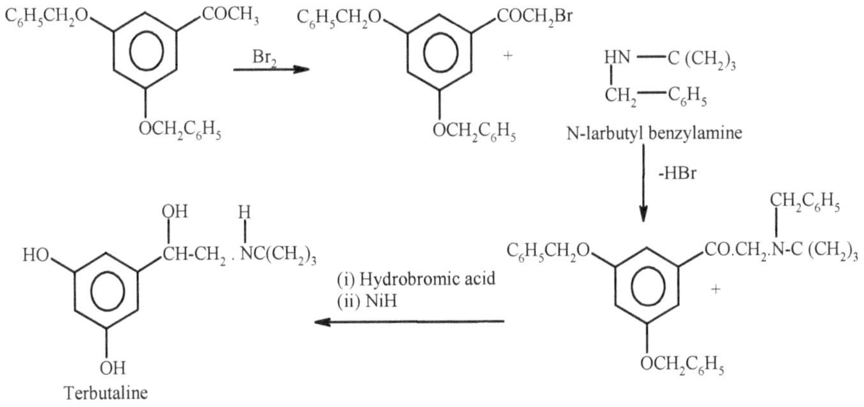

N-larbutyl benzylamine

-HBr

(i) Hydrobromic acid
(ii) NiH

Terbutaline

Salbutamol (Albuterol)

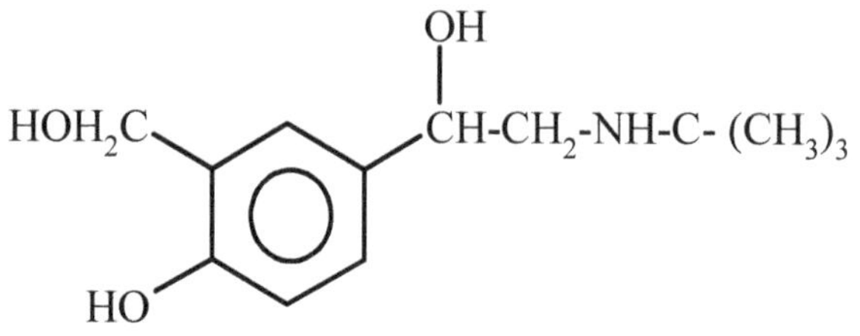
OH
HOH2C
CH-CH2-NH-C- (CH3)3
HO

4-Hydroxy -3-hydroxy methyl α - [(ict - butyl amino) methyl] benzyl alcohol

Synthesis

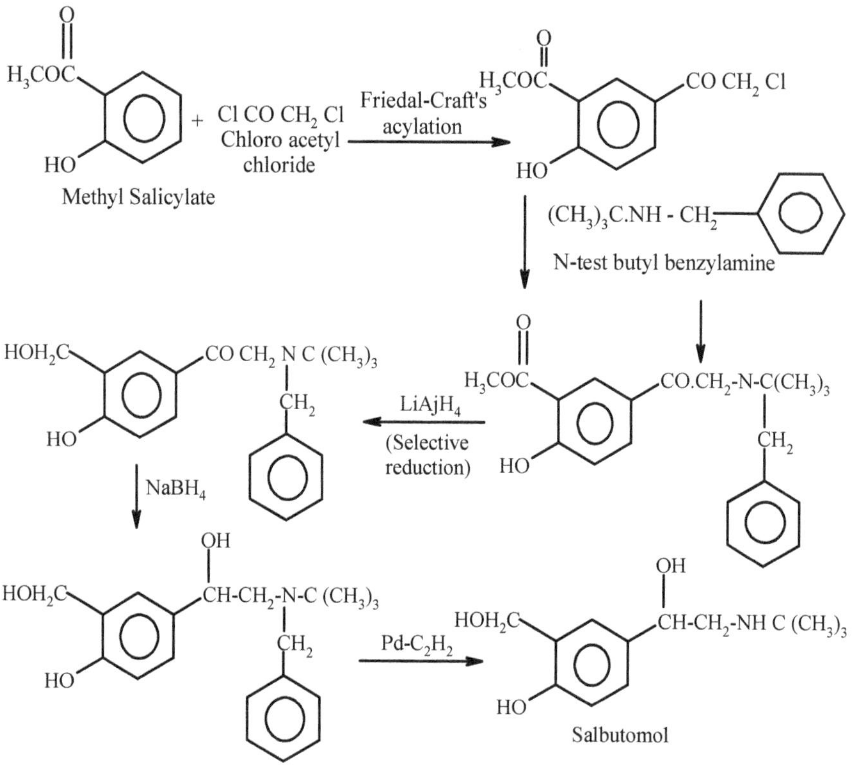
O
H3COC
HO
Methyl Salicylate
+ Cl CO CH2 Cl
Chloro acetyl chloride
Friedal-Craft's acylation
O
H3COC
CO CH2 Cl
HO
(CH3)3C.NH - CH2
N-test butyl benzylamine
O
H3COC
CO.CH2-N-C(CH3)3
CH2
HO
LiAjH4
(Selective reduction)
HOH2C
CO CH2 N C (CH3)3
CH2
HO
NaBH4
OH
HOH2C
CH-CH2-N-C (CH3)3
CH2
HO
Pd-C2H2
OH
HOH2C
CH-CH2-NH C (CH3)3
HO
Salbutomol

D. Both a and b-Adrenergic Receptor Agonist

Dobutamine

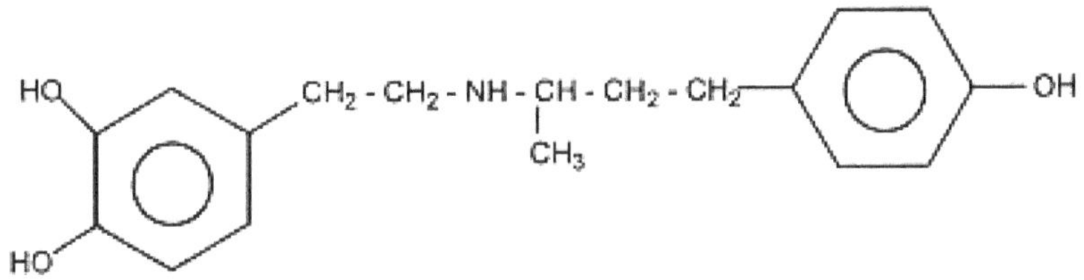
HO
HO
CH2 - CH2 - NH - CH - CH2 - CH2
CH3
OH

N – [1 – Methyl – 3 – (4 – hydroxyphenyl) propyl] 3, 4 - dihydroxy phenyl ethyl amine

2. Indirect acting Adrenergic Agonists

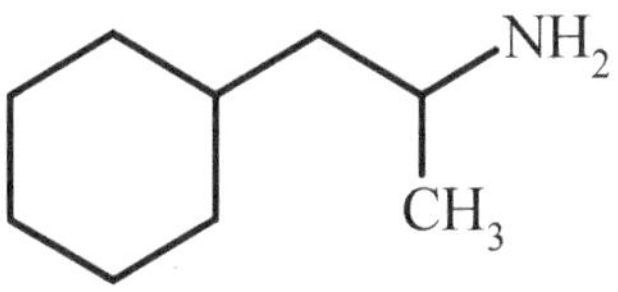
NH2
CH3

Propylhexedrine

3. Dual acting Adrenergic Agonists

Ephedrine

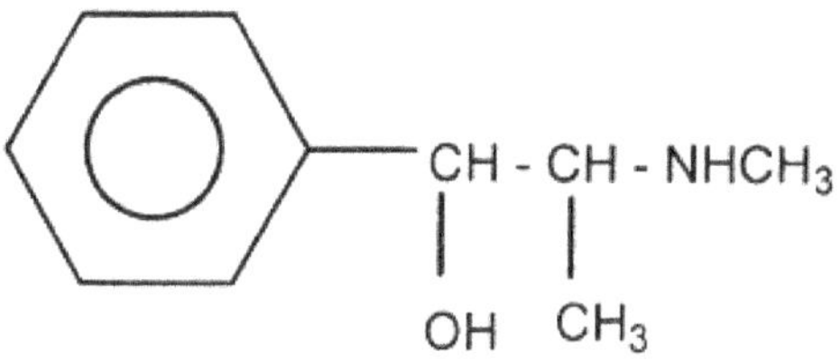
CH - CH - NHCH3
OH CH3

(1R,2S) - 2-Methyl amino -1-phenyl propan -1-ol

Synthesis

Synthesis of Amphetamine

SAR for Adrenergic Agonists

General structure

SAR for sympathomimetic drugs are discussed in the following category.

A. Substitution in the Phenyl ring system

1. The receptor selectivity of the drugs depends on the substituent at aromatic ring, a,b- Carbon and Amino group.

2. The maximal sympathomimetic activity shown by substitution of hydroxyl group in M and P position of aromatic ring.

3. The amino group should be separated from the aromatic ring by two carbon atoms for optimal activity.

4. The naturally occurring Nor Adrenaline has 3', 4'-dihydroxy benzene ring active at both a and b-Adrenergic receptors.

5. The dihydroxy substitution at 3', 4' position (Ex-Metaproterinol) gives good oral activity and selectivity for b2-Adrenergic receptors.

6. Their substituents like 3'-hydroxy methyl (albuterol), 3'-trifluoro methyl, 4' amino, 5' chloro (Mebuterol) have good oral activity.

7. One hydrogen bonding group is essential at the 4' position for beta activity and 3'- OH substitution for Alpha activity.

B. Substitutions at Nitrogen (Amino group)

1. The presence of amino group is important for direct agonist activity.

2. Primary and secondary amines are more potent direct acting agonist than tertiary amine.

3. Size in alkyl group of nitrogen increases, a-receptor agonist activity decreases and beta-receptor agonist activity increases (Ex-Isoproterenol).

4. N-substitution also provides selectivity for different b-receptor sub types. Large t-butyl group have selectivity to b$_2$ receptor (Ex-Colterol). Ritordrine with large P-Hydroxy phenyl ethyl substitution is a selective b$_2$ agonist.

5. Nitrogen in the part of heterocyclic ring such as imidazoline possess anti hypertensive property.

C. Substitutions on Carbon in the side chain

1. There are two carbon atoms a and b to nitrogen function. Small alkyl groups such as methyl or ethyl present in the

a-carbon, an ethyl group at this position diminish the a-activity for more than b-activity.

2. The presence of a-alkyl group increases the duration of action by making the compound resistant to metabolic deamination by MAO.

3. Maximal direct activity in a-methyl nor adrenaline in the erythro enantiomers. b-carbon has hydroxyl group in the (R) absolute configuration for   maximal direct activity.

Mechanism of Action

1. *For Directly acting Sympathomimetics*: They act through complexation with specific receptors. For the activation of b-receptor phenolic hydroxy group in meta position of aromatic group and an alcoholic hydroxyl group in b-position of side chain and an amine with bulky group.

2. *For InDirectly acting Sympathomimetics*: They act either by releasing catecholamines mainly norepinephrine from storage granules in the sympathetic nerve terminals or through inhibition of nor adrenaline uptake at the neuronal membrane.

Anti- Adrenergic Drugs

Adrenergic Antagonists

Sympatholytics (or) Adrenergic Receptor Blocking Agents

Adrenergic antagonists are drugs that reduce the delivery of catecholamines to the adrenergic receptors by disrupting catecholamine synthesis, storage or release. They abolish the response to stimulation of sympathetic nerves.

These agents competitively antagonize the effect of the catecholamines at α and β adrenergic receptors.

Classification

I.   **α – Adrenergic blocking agents**

1. Imidazolines – Tolazoline, Phentolamine

2. Beta Halo Alkylamines – Phenoxy benzamine, Dibenamine

3. Quinazolines – Prazosin, Terazosin, Doxazosin

4. Ergot Alkaloids – Ergotamine, Ergosin, Ergocrystin, Ergocriptine

5. Miscellaneous – Yohimbin, Methy sergide

II. β – Adrenergic blocking agents

1. Aryl ethanolamines – Isoproterenol, pronethalol, Dichloro isoproterenol

2. Aryloxy propanolamines – Propranolol, Practalol, Metaprolol, Acebutolol, Atenolol, Betaxolol, Bisoprolol, Esmolol.

III. Both α and β – Adrenergic blocking agents

Labetalol, Carvedilol.

I. α – Adrenergic blocking agents

A. Tolazoline

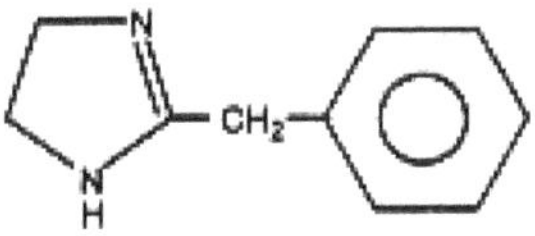

2-Benzyl - 2- imidazoline

Synthesis

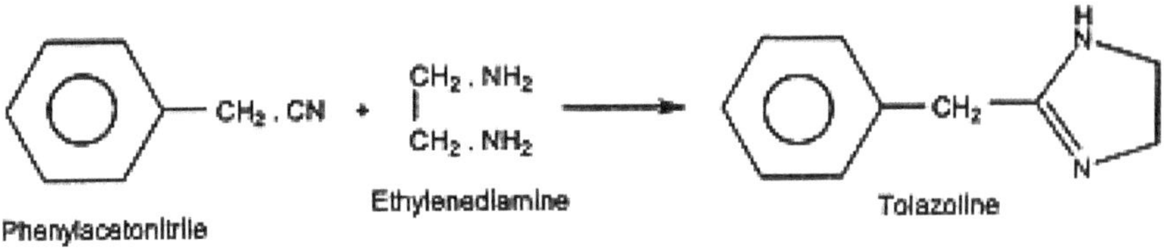

Phenylacetonitrile

Ethylenediamine

Tolazoline

Phentolamine

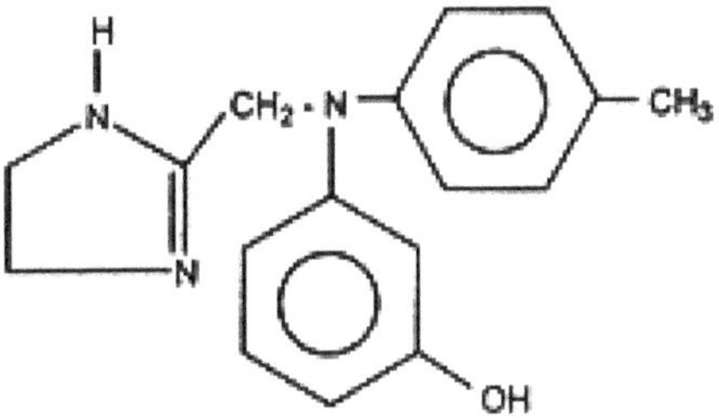

3-{[(4, 5-Dihydro-1H-imidazol-2-yl) methyl] (4-methyl phenyl) – amino} phenol

Ergot alkaloids

Name	R_1	R_2
Ergotamine	$- CH_3$	$- CH_2\, C_6\, H_5$
Ergocristine	$- CH\,(CH_3)_2$	$- CH_2\, C_6\, H_5$
Ergocryptine	$- CH\,(CH_3)_2$	$- CH_2\, CH(CH_3)_2$
Ergocornine	$- CH\,(CH_3)_2$	$- CH\,(CH_3)_2$

II. β – Adrenergic blocking agents

These drugs block the effects of Endogeneous and exogeneous catecholamines. These drugs slow the heart rate and decrease the force of contraction. They competitively inhibit β – Adrenergic receptors. These are also used in the treatment of hypertension, arrhythmiasis, coronary artery disease and open angle glaucoma

Propranolol

1-(Isopropyl amino) -3- (1-naphthyloxy) -2- propanol

Dichloroisoproterenol

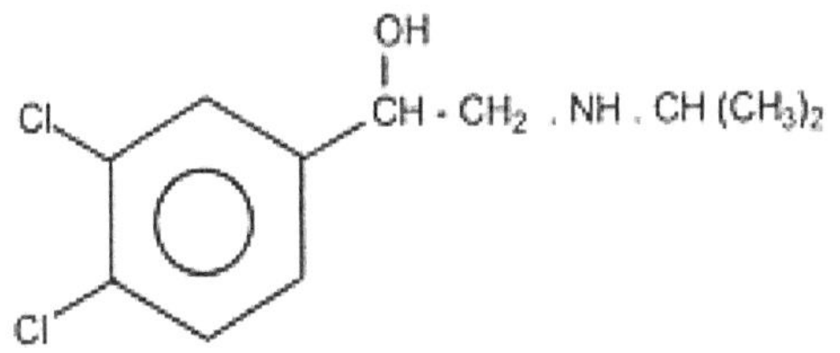

3, 4 - Dichloro - α - [(isopropyl amino) methyl] benzyl alcohol.

Metoprolol

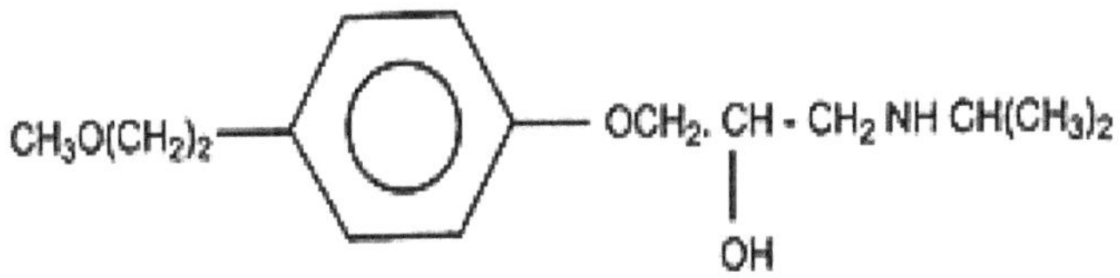

1 - [4 - (2-Methoxyethyl) phenoxy] - 3 - [(1- methylethyl) amino] - 2 - propanol

Atenolol

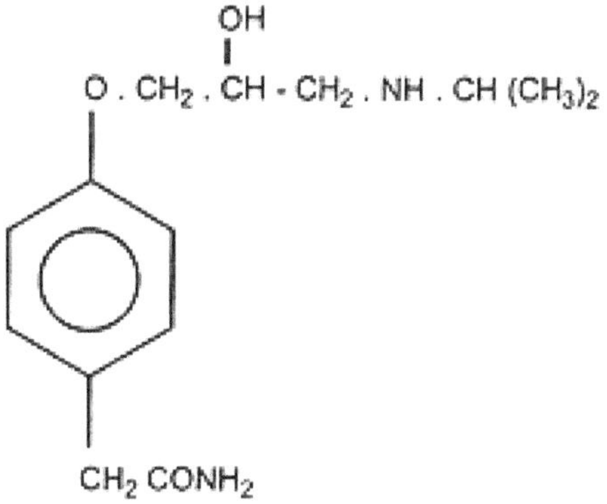

4-{2-Hydroxy-3-[(1-methylethyl) amino) propoxy} benzeneacetamide

III.   Both α and β – Adrenergic blocking agents

Labetolol

$$OH$$
CH - CH$_2$. NH . CH . CH$_2$. CH$_2$
CH$_3$
HO
CONH$_2$

5-{1-Hydroxy -2- [(1-methyl -3-phenylpropyl) amino} ethyl] salicylamide.

SAR for Beta Blockers

$$Ar - O - CH_2 - CH - CH_2 \; NH . R$$
$$OH$$

General structure

1. The O-CH$_2$ group between aromatic ring and the ethylamino side chain is responsible for the antagonistic property.

2. Replacement of catechol hydroxyl group with chlorine or phenyl ring retains the beta blocking activity.

3. N,N- di substitution decrease beta blocking activity. Activity is maintained when phenylethyl, hydroxyl phenyl ethyl or methoxy phenyl ethyl groups are added to amine as a part of molecule.

4. The two carbon side chain is essential for the activity.

5. Nitrogen atom should be of secondary amine for optimum beta blocking activity.

6. The carbon side chain having hydroxyl group must be S- configuration for optimum affinity to beta receptor. (Ex- Levobunolol, Timolol)

7. The aryloxy propanolamines are more potent than aryl ethanolamines.

8. Replacement of ethereal oxygen in aryloxy propanolamines with S, CH$_2$ or N-CH$_3$ is decreased the beta blocking activity.

9. The most effective substituents at amino group is isopropyl and tertiary butyl group.

10. The aromatic portion of the molecules could be varied with good activity.

11. Converting the aromatic portion to phenanthrene or anthracene decrease the activity.

12. Cyclic alkyl substituents are better than corresponding open chain substituents at nitrogen atom of amine.

13. Alpha methyl group at side chain decrease activity.

Mechanism of Action

1. These drugs competitively inhibit the adrenergic receptors.

2. Beta antagonists are invariably employed in the treatment of essential hypertension and cause an effective decrease in BP by exerting direct effect on heart and blood vessels, minimizing sympathetic outflow from CNS and affecting the rennin-angiotensin- aldosterone system.

3. Some drugs like propranolol precipitate an asthmatic attack by antagonizing beta-2 receptors in bronchial smooth muscle and give rise to sudden contraction of bronchial smooth muscle.

Neuromuscular Blocking Agents

Neuromuscular-blocking drugs block neuromuscular transmission at the neuromuscular junction, causing paralysis of the affected skeletal muscles. This is accomplished either by acting presynaptically via the inhibition of acetylcholine (ACh) synthesis or release, or by acting postsynaptically at the acetylcholine receptor. While there are drugs that act presynaptically (such as botulin toxin and tetrodotoxin), the clinically-relevant drugs work postsynaptically.

Quaternary ammonium muscle relaxants are quaternary ammonium salts used as drugs for muscle relaxation, most commonly in anasthesia. It is necessary to prevent spontaneous movement of muscle during surgical operations. Muscle relaxants inhibit neuron transmission to muscle by blocking the nicotinic acetylcholine receptor. What they have in common and is necessary for their effect is having quaternary ammonium groups, usually two. Some of them are found in nature and others are synthesized molecules.

Neuromuscular Blocking Agents

An agent which blocks the transmission of Ach at the motor end place are called neuromuscular blocking agents. They are used in surgical anesthesia as adjuvant to relax the skeletal muscle.

Classification

1. **Natural Compounds:** Eg: Tubocurarine chloride, Metocurarine iodide, Pancuronium bromide.

2. **Synthetic Compounds:** Eg: Gallamine triethiodide, Decamethonium bromide, Pipecuranium bromide, Vecuronium bromide

Classification

These drugs fall into two groups:

- **Non-depolarizing blocking agents**: These agents constitute the majority of the clinically-relevant neuromuscular blockers. They act by blocking the binding of ACh to its receptors, and in some cases, they also directly block the ionotropic activity of the ACh receptors

- **Depolarizing blocking agents**: These agents act by depolarizing the plasma membrane of the skeletal muscle fiber. This persistent depolarization makes the muscle fiber resistant to further stimulation by ACh.

SAR for Neuro Muscular Blockers

1. The drugs have quarternary ammonium group for good activity.
2. The type of alkyl group present in quarternary ammonium group determines the charge distribution and binding characters.
3. Non depolarizing drugs are bulky and more rigid than depolarizing drugs.
4. As the strerric hindrance to receptor increases, the potency decreases. L-tubocurarine is less potent than d-tubocurarine.

5. The depolarizing agents (Eg-Decamethonium) have a more flexible structure that enable bond rotation.

6. The distance between quarternary ammonium groups can vary up to the limit of maximal bond distance usually 1.0 ± 0.1 nm.

Mechanism of Action

Neuro muscular blocking agents can block the neuro muscular transmission by

- Inhibiting acetyl choline synthesis

- Inhibiting Ach release and inhibit calcium entry have neuromuscular block. Interfering with the post synoptic action of Ach.

- Non depolarizing blocking agents act by competitive antagonism at Ach receptos of the end plate and these largely accounts for their action.

Gallamine (as **gallamine triethiodide**) is a non-depolarising muscle relaxant also known under the trade name **Flaxedil.** It acts by combining with the cholinergic receptor sites in muscle and competitively blocking the transmitter action of acetylcholine. Gallamine has a parasympatholytic effect on the cardiac vagus nerve which causes tachycardia and occasionally hypertension. Very high doses cause histamine release.

Gallamine is commonly used to stabilize muscle contractions during surgical procedures

Gallamine Triethiodide

$$\left[\begin{array}{c} O\ CH_2.CH_2.\overset{\oplus}{N}\text{-}(CH_2\ CH_3)_3 \\ O\ CH_2.CH_2.\overset{\oplus}{N}\text{-}(CH_2\ CH_3)_3 \\ O\ CH_2.CH_2.\overset{\oplus}{N}\text{-}(CH_2\ CH_3)_3 \end{array} \right] \quad 3\ \overset{\ominus}{I}$$

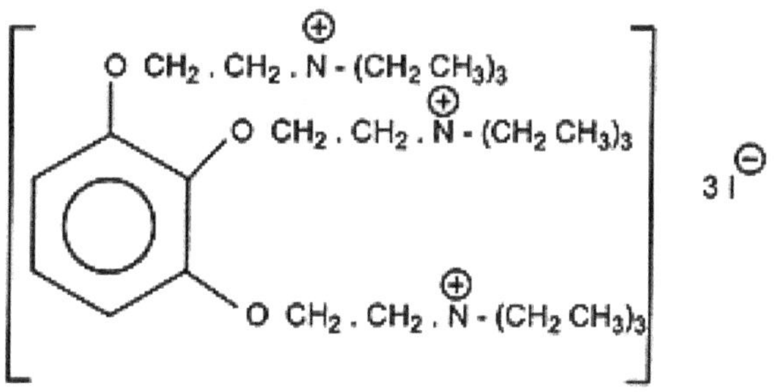

[2 – phenenyl - tris (oxyethylene)] tris (triethyl ammonium) triiodide

Synthesis

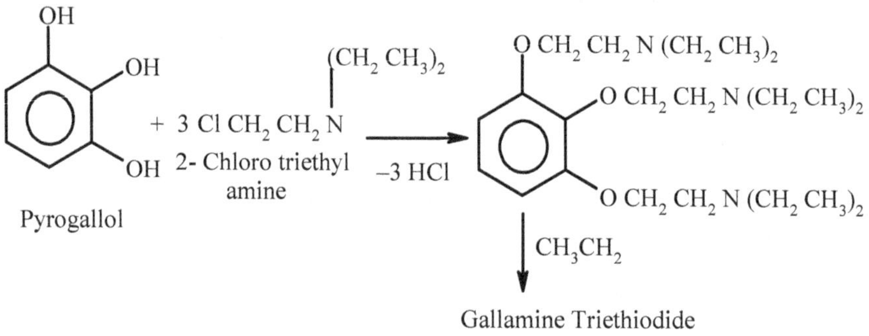
OH
OH
OH
Pyrogallol
(CH2 CH3)2
+ 3 Cl CH2 CH2 N
2- Chloro triethyl amine
-3 HCl
O CH2 CH2 N (CH2 CH3)2
O CH2 CH2 N (CH2 CH3)2
O CH2 CH2 N (CH2 CH3)2
CH3CH2
Gallamine Triethiodide

Mephenesin

Mephenesin is a centrally acting muscle relaxant. Once used in association with pentobarbital for anesthesia, but no longer used because of its many side-effects, including venous thrombosis and hemolysis.

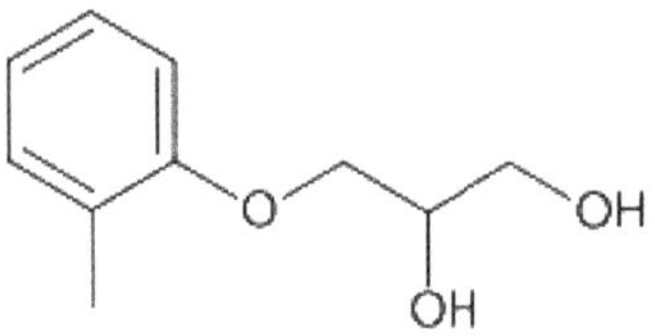
O
OH
OH

Pancuronium

Pancuronium is a chemical compound, used in medicine as the bromide salt pancuronium bromide. It has the brand name Pavulon (Organon International). It is a muscle relaxant with various purposes. It is one of the drugs administered during a lethal injection in the United States

Mode of action

Pancuronium is a typical non-depolarizing curare-mimetic muscle relaxant. It acts as a competitive acetylcholine antagonist on neuromuscular junctions, displacing acetylcholine (hence competitive) from its post-synaptic nicotinic acetylcholine receptors. It is (unlike suxamethonium) a non-depolarizing agent, which means that it causes no spontaneous depolarizations upon association with the nicotinic receptor in neuromuscular junction, thus producing no muscle fasciculations upon administration. Despite being a steroid, pancuronium has no hormonal activity. It exerts slight vagolytic activity (i.e. diminishing activity of the vagus nerve) and no ganglioplegic (i.e. blocking ganglions) activity. Pancuronium is a very potent muscle relaxant/curaremimetic. The ED95 (i.e. a dose causing a 95% reduction in muscle activity) is only 60 µg/kg body weight administered intravenously. Muscle relaxation suitable for intubation sets in about 90–120 seconds after administration of the drug. Full muscle paralysis for major surgery is achieved about 2–4 minutes after application. Clinical effects (muscle activity lower than 25% of physiological) last for about 100 minutes. The time needed for full (over 90% muscle activity) recovery after single administration is about 120–180 minutes in healthy adults, but can be protracted to more hours in poor health subjects and when concomitantly administered with other long-acting anesthetics (e.g. some opioids, barbiturates, inhalation anesthetics). The effects of pancuronium can be at least partially reversed by anticholinesterasics, such as neostigmine, pyridostigmine and edrophonium.

[(2*S*,3*S*,5*S*,8*R*,9*S*,10*S*,13*S*,14*S*,16*S*,17*R*)-17-acetyloxy-10,13-dimethyl-2,16-bis(1-methyl-3,4,5,6-tetrahydro-2*H*-pyridin-1-yl)-2,3,4,5,6,7,8,9,11,12,14,15,16,17-tetradecahydro-1*H* cyclopenta[*a*]phenanthren-3-yl] acetate

Synthesis

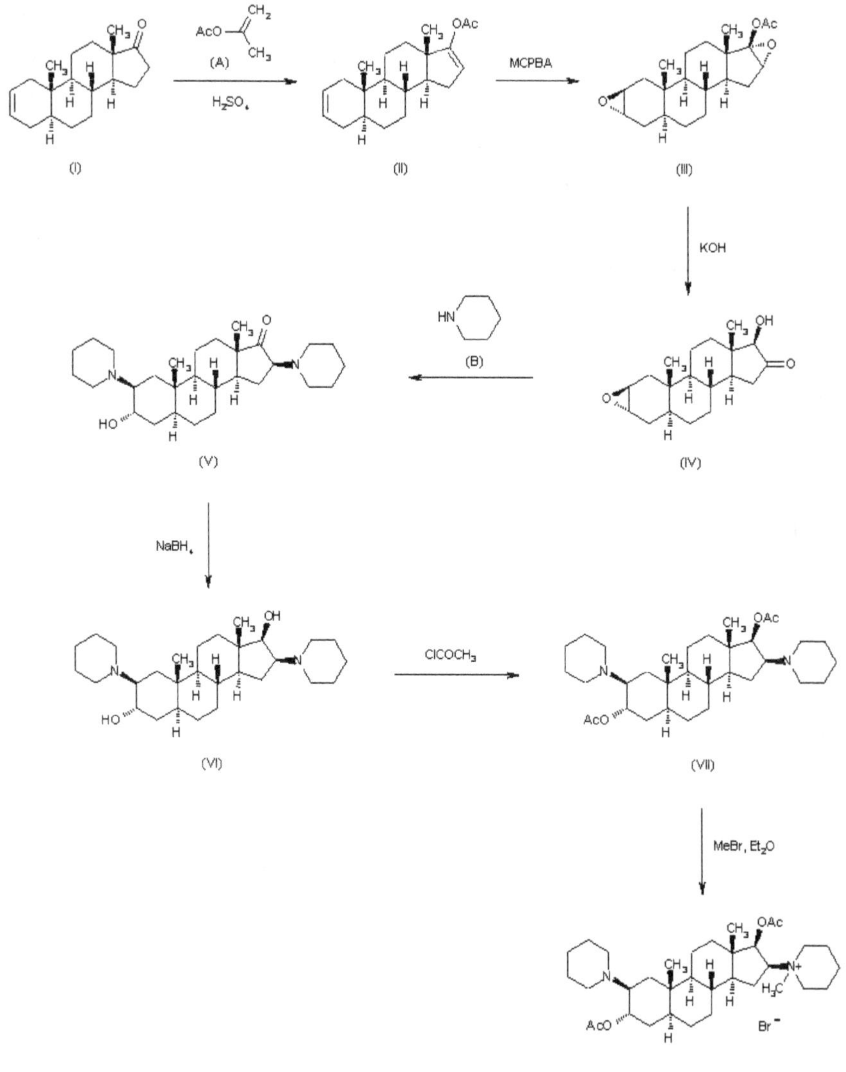

Drugs used in the Treatment of Alziemer's Disease

Alzheimer's disease (AD), also called **Alzheimer disease, Senile Dementia of the Alzheimer Type (SDAT)** or simply **Alzheimer's** is the most common form of dementia. This incurable, degenerative, and terminal disease was first described by German psychiatrist and neuropathologist Alois Alzheimer in 1906 and was named after him. Generally, it is diagnosed in people over 65 years of age, although the less-prevalent early-onset Alzheimer's can occur much earlier. As of September 2009, this number is reported to be 35 million-plus worldwide. The prevalence of Alzheimer's is thought to reach approximately 107 million people by 2050.

Although the course of Alzheimer's disease is unique for every individual, there are many common symptoms. The earliest observable symptoms are often mistakenly thought to be 'age-related' concerns, or manifestations of stress. In the early stages, the most commonly recognised symptom is memory loss, such as difficulty in remembering recently learned facts. When a doctor or physician has been notified, and AD is suspected, the diagnosis is usually confirmed with behavioural assessments and cognitive tests, often followed by a brain scan if available.

As the disease advances, symptoms include confusion, irritability and aggression, mood swings, language breakdown, long-term memory loss, and the general withdrawal of the sufferer as their senses decline. Gradually, bodily functions are lost, ultimately leading to death. Individual prognosis is difficult to assess, as the duration of the disease varies. AD develops for an indeterminate period of time before becoming fully apparent, and it can progress undiagnosed for years. The mean life expectancy following diagnosis is approximately seven years. Fewer than three percent of individuals live more than fourteen years after diagnosis.

The cause and progression of Alzheimer's disease are not well understood. Research indicates that the disease is associated with plaques and tangles in the brain. Currently used treatments offer a small symptomatic benefit; no treatments to delay or halt the progression of the disease are as yet available. As of 2008, more than 500 clinical trials have been conducted for identification of a possible treatment for AD, but it is unknown if any of the tested intervention strategies will show promising results. A number of non-invasive, life-style habits have been suggested for the prevention of Alzheimer's disease, but there is a lack of adequate evidence for a link between these recommendations and reduced degeneration. Mental stimulation, exercise, and a balanced diet are suggested, as both a possible prevention and a sensible way of managing the disease.

Because AD cannot be cured and is degenerative, management of patients is essential. The role of the main caregiver is often taken by the spouse or a close relative. Alzheimer's disease is known for placing a great burden on caregivers; the pressures can be wide-ranging, involving social.

Drugs: Donepezil, Rivastigmine and Galantamine

Donepezil (also misspelled **donezepil**), marketed under the trade name **Aricept** by its developer Eisai and partner Pfizer, is a centrally acting reversible acetylcholinesterase inhibitor. Its main therapeutic use is in the treatment of Alzheimer's disease where it is used to increase cortical acetylcholine. Its binding to the acetylcholinesterase can be seen at *Proteopedia 1eve*. It has an oral bioavailability of 100% and easily crosses the blood-brain barrier. Because it has a half life of about 70 hours, it can be taken once a day. Initial dose is 5 mg per day, which can be increased to 10 mg per day after an adjustment period of at least 4 weeks.

IUPAC Name: (*RS*)-2-[(1-benzyl-4-piperidyl) methyl]- 5,6-dimethoxy-2,3-dihydroinden-1-one

Mechanism of action: Inhibits acetylcholinesterase, increasing the amount of acetylcholine in synapses and potentially enhancing brain cholinergic neurotransmission.

Studies in Alzheimer's Disease

Currently, there is no definitive proof that use of donepezil or other similar agents alters the course or progression of Alzheimer's disease. However, 6-12 month controlled studies have shown modest benefits in cognition and/or behavior, such as one published in 1999. Pilot studies have reported that donepezil therapy may potentially have effects on markers of disease progression, such as hippocampal volume. Therefore, many neurologists, psychiatrists, and primary care physicians use donepezil in patients with Alzheimer's disease. In 2005, the UK National Institute for Clinical Excellence (NICE) withdrew its recommendation for use of the drug for mild-to-moderate AD, on the basis that there is no significant improvement in functional outcome; of quality of life or of behavioral symptoms. However, NICE revised its guidelines to suggest that donepezil be used in moderate stage patients for whom the evidence is strongest. It is currently not licensed for Alzheimer's disease in the UK at any other stage. While the drug is currently indicated for mild to moderate Alzheimer's, there is also evidence from 2 trials that it may be effective for moderate to severe disease. An example of this is a Karolinska Institute paper published in *The Lancet* in early 2006, which states that donepezil improves cognitive function even in patients with severe Alzheimer's disease symptoms.

Use in the General Population

In July 2002, a pilot study, reported that donepezil improves the memory of aging pilots. The researchers trained pilots in a flight simulator to perform specific maneuvers and to respond to emergencies that developed during their mock flight, after giving half the pilots donepezil and half a placebo. One month later they retested the pilots and found that those who had taken the donepezil remembered their training better, as shown by improved performance

Rivastigmine

IUPAC NAME : *S*)-*N*-Ethyl- *N*-methyl- 3-[1-(dimethylamino)ethyl]-phenyl carbamate

Rivastigmine (sold under the trade name **Exelon**) is a parasympathomimetic or cholinergic agent for the treatment of mild to moderate dementia of the Alzheimer's type and dementia due to Parkinson's disease. The drug can be administered orally or via a transdermal patch; the latter form reduces the prevalence of side effects, which typicaly include nausea and vomiting. The drug is eliminated through the urine, and appears to have relatively few drug-drug interactions

Rivastigmine is also a cholinesterase inhibitor (ChEI)

History

Rivastigmine was developed by Novartis, and has been available in capsule and liquid formulations since 1997. In 2006, it became the first product approved globally for the treatment of mild to moderate dementia associated with Parkinson's Disease; and in 2007 the rivastigmine transdermal patch became the first patch treatment for dementia.

Administration

Rivastigmine tartrate is a white to off-white fine crystalline powder that is both lipophilic (soluble in fats) and hydrophilic (soluble in water). Like other cholinesterase inhibitors, it requires doses to be increased gradually over several weeks; this is usually referred to as the *titration phase*. Oral doses of rivastigmine should be titrated with a 3 mg per day increment every 2 to 4 weeks.

Rivastigmine is classified as Pregnancy category B, with insufficient data on risks associated with breastfeeding. In cases of overdose, atropine is used to reverse bradycardia. Dialysis is ineffective due to the drug's half-life

Galantamine

((4a*S*,6*R*,8a*S*)- 5,6,9,10,11,12- hexahydro- 3-methoxy- 11-methyl-4a*H*-[1]benzofuro [3a,3,2-*ef*] [2] benzazepin- 6-ol)

Galantamine (Nivalin, Razadyne, Razadyne ER, Reminyl) is a chemical used for the treatment of mild to moderate Alzheimer's disease and various memory impairments. It is an alkaloid that is obtained synthetically or from the bulbs and flowers of the Caucasian snowdrop (Voronov's snowdrop), *Galanthus woronowii* (Amaryllidaceae) and related genera like *Narcissus* (daffodil), *Leucojum* (snowflake) and *Lycoris* including *Lycoris radiata* (Red Spider Lily). The active ingredient was isolated by prof. Paskov 1959 (Sopharma, Bulgaria) from a species tradionally used as a popular medicine in Eastern Europe and thus the idea for developing a medicine from these species seems to be based on the local use (ie, an ethnobotany-driven drug discovery). It has been used for decades in Eastern Europe esp. in the symptomatic treatment of polio (poliomyelitis) and was later developed by Janssen Pharmaceutica into an Alzheimer medication. In the US it has been sold as a dietary supplement for memory and dream support prior to being approved as a drug by the FDA.

Pharmacology

Galantamine in its pure form is a white powder. Galantamine is a competitive and reversible cholinesterase inhibitor. It is believed it works by enhancing cholinergic function by increasing the concentration of acetylcholine in the brain. The atomic resolution 3D structure of the complex of galantamine and its target, acetylcholinesterase, was determined by X-ray crystallography in 1999 (PDB code: 1DX6; see complex).[3] There is no evidence that galantamine alters the course of the underlying dementing process. Galantamine has also shown activity in modulating the nicotinic cholinergic receptors to increase acetylcholine release.

Metabolism

The major route of metabolism for galantamine is through the liver, this accounts for approximately 75% of the total metabolism of galantamine. Hepatic cytochrome P450 (CYP) isoenzymes are the active enzymes for this metabolic route. *In vitro* studies have shown that CYP2D6 and CYP3A4 are involved in galantamine metabolism.

For Razadyne ER (the once-a-day formulation), CYP2D6 poor metabolizers had drug exposures that were approximately 50% higher than for extensive metabolizers. About 7% of the population has this genetic mutation, however because the drug is individually titrated to tolerability, no specific dosage adjustment is necessary for this population.

Local Anaesthetics

A **local anesthetic** is a drug that causes reversible local anesthesia and a loss of nociception. When it is used on specific nerve pathways (nerve block), effects such as analgesia (loss of pain sensation) and paralysis (loss of muscle power) can be achieved.

Clinical local anesthetics belong to one of two classes: aminoamide and aminoester local anesthetics. Synthetic local anesthetics are structurally related to cocaine. They differ from cocaine mainly in that they have no abuse potential and do not act on the sympathoadrenergic system, i.e. they do not produce hypertension or local vasoconstriction, with the exception of Ropivacaine and Mepivacaine that do produce weak vasoconstriction.

Local anesthetics vary in their pharmacological properties and they are used in various techniques of local anesthesia such as:

- Topical anesthesia (surface)
- Infiltration
- Plexus block
- Epidural (extradural) block
- Spinal anesthesia (subarachnoid block)

The local anesthetic lidocaine (lignocaine) is also used as a Class Ib antiarrhythmic drug.

Classification

1. Naturally occurring local anaesthetic: Ex- Cocaine (ester).

2. Esters:

    (a) *P-aminobenzoic acid derivative (PABA):* Benzocaine, Procaine, Tetracaine, Butacaine, orthocaine, Benoxinate.

    (b) *Esters of benzoic acid:* Meprylcaine, Cyclomethycaine, (propoxycaine), Hexylcaine, Piperocaine.

3. Amides or Anilides: Lignocaine, Prilocaine, Mepivacaine, Bipivacaine, Pyrrocaine, Etidocaine, Diperodone.

4. Piperidine or Tropane derivatives: Alpha-Eucaine, Benzamine, Euphthalmin.

5. Quinoline derivatives: Dibucaine (Chincocaine)

6. Isoquinoline derivative: Dimethisoquine.

7. Miscelleneous: Phenacaine (Amidine), Pramoxine, Euginol, benzyl alcohol, Phenol, Dyclomine, Saligenin.

Piperocaine (Metycaine)

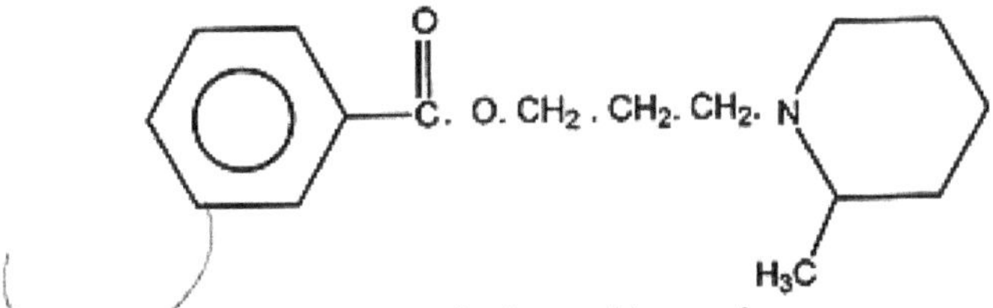

3-(2-Methyl piperidino) propyl benzoate

Phenacaine

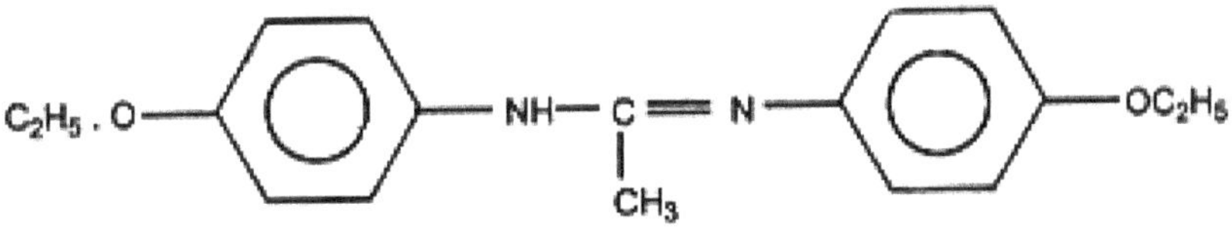

N,N' - bis (4-ethoxy phenyl) ethanimidamide

Lidocaine

Lidocaine or **lignocaine** is a common local anesthetic and antiarrhythmic drug. Lidocaine is used topically to relieve itching, burning and pain from skin inflammations, injected as a dental anesthetic or as a local anesthetic for minor surgery.

Preparation

Lidocaine may be prepared in two steps by the reaction of 2, 6-xylidine with chloroacetyl chloride, followed by the reaction with diethylamine

A. Lidocaine (Xylocaine, Lignocaine)

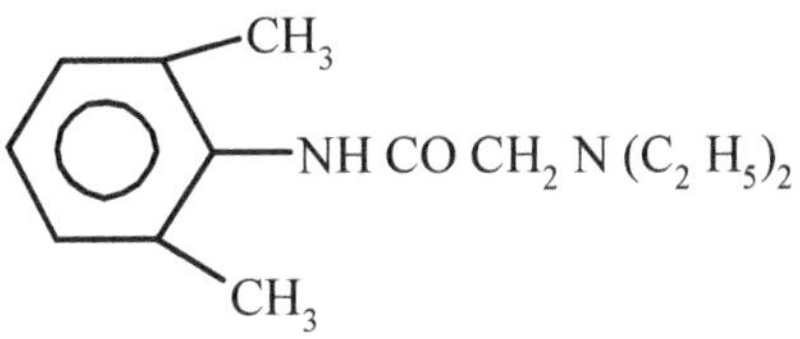

N,N - (Diethyl amino aceto) - 2, 6 - xylidine

Synthesis

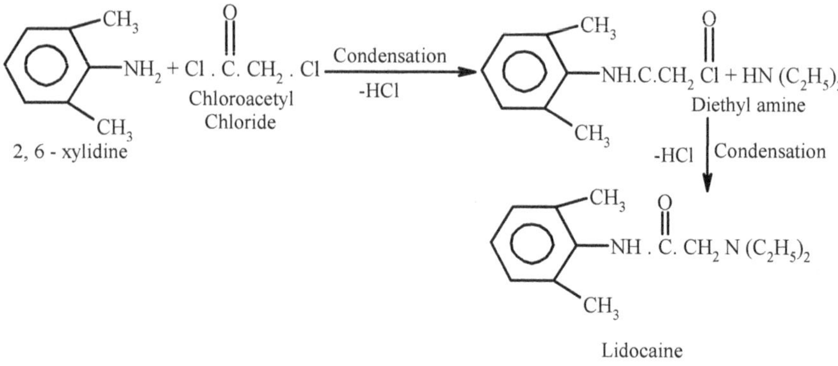

2, 6 - xylidine + Cl.C.CH$_2$.Cl (Chloroacetyl Chloride) $\xrightarrow[\text{-HCl}]{\text{Condensation}}$ ... -NH.C.CH$_2$ Cl + HN (C$_2$H$_5$)$_2$ (Diethyl amine) $\xrightarrow[\text{Condensation}]{\text{-HCl}}$ ... -NH . C. CH$_2$ N (C$_2$H$_5$)$_2$

Lidocaine

Benzocaine

Benzocaine is a local anesthetic commonly used as a topical pain reliever. It is the active ingredient in many over-the-counter anesthetic ointments (e.g. products for oral ulcers of Anbesol by Wyeth, Kank+a by Blistex, Orabase B and Orajel by Del Pharmaceuticals, and Ultracare by Ultradent). It is also combined with antipyrine to form A/B Otic Drops, (Brand name **Auralgan**) to relieve earpain and remove cerumen.

Chemical Properties

Benzocaine is the ethyl ester of p-aminobenzoic acid (PABA); it can be prepared from PABA and ethanol by Fischer esterification or via the reduction of ethyl p-nitrobenzoate. Benzocaine is sparingly soluble in water, though is more soluble in dilute acids and very soluble in ethanol, chloroform and ethyl ether. The melting point of benzocaine is 88-92 degrees Celsius and the boiling point is 172 degrees Celsius. The density of benzocaine is 1.17 g/cm^3.

History

Benzocaine was first synthesised by a German chemical firm named Ritsert, in the town of Eberbach.

Mechanism of Action

Pain is caused by the stimulation of free nerve endings. When the nerve endings are stimulated, sodium enters the neuron, causing depolarization of the nerve and subsequent initiation of an action potential. The action potential is propagated down the nerve toward the central nervous system, which interprets this as pain. Esters of PABA work as a chemical barrier, stopping the sodium entering the nerve ending.

Synthesis of Benzocaine

In this experiment, the synthesis of benzocaine (2) is carried out by the acid-catalyzed esterification of 4-aminobenzoic (1) acid with ethanol:

SAR for Local Anaesthetic Containing Ester Linkage

General Structure

Aryl – CO – X - Amino alkyl Chain

1. The aryl radical attached directly to the carbonyl group enhances local anaesthetic activity. It is lipophilic centre of compound.

2. Alicyclic and aryl aliphatic carboxylic acid esters are also active local anaesthetics.

3. The compounds containing aryl-vinyl group (Ar-CH = CH -) does not having local anaesthetic activity, because of the mesomeric effect of aryl radical does not extend to carbonyl group.

4. The aryl substituents such as alkoxy, amino and alkyl amino groups at ortho or para position increases electron density of carbonyl oxygen enhances the activity.

5. The number of methylene groups is substituted to aryl moiety; the maximum activity is achieved for the C4 to C6 homologues.

6. The bridge X may be carbon, oxygen, nitrogen or sulphur. The nature of X affects duration of action and relative toxicity. The conduction anaesthetic potency decreases in the order of S, O, C and N.

7. In general amino alkyl group is not necessary for activity, but it is used to form water soluble salts. Example : Benzocaine.

8. Local anaesthetic activity improves if the aryl lipophilic center has electron donor substitution but decreases with electron acceptor substituents.

SAR for Local Anaesthetic Containing an Amide Linkage

1. The alkyl substitution ($-CH_3$) in aryl group at ortho or para position enhances the activity by providing steric hindrance to the hydrolysis of amide linkage and contributes lipid solubility.

2. In general X may be carbon (Isogramine), oxygen (lidocaine) (or) nitrogen (Phenacaine) for good activity.

3. The relative activity of amino alkyl group is similar to the ester linkage containing compounds.

Mechanism of Action

1. The main site of local anesthetic action is on the cell membrane. They do not produce any effect on the intracellular fluid (Axoplasm). Drugs produce local anesthetic effect by inhibiting membrane conductance and formation of action potential by either fully or partially blocking the sodium ion channels.

2. If sufficient numbers of sodium channels are blocked there would be no significant charges in membrane potential and so the conduction of an action potential along the neuron would be prevented.

3. Blocking of conduction would automatically prevent the release of neuro transmitter at the phesynoptic site.

4. Increasing the concentration of calcium ions of the extra cellular fluid may enhance or reduce the activity by affecting the opening of sodium channels.

5. Local anesthetic containing both lipophilic and hydrophilic groups may penetrate the excitable cells and decrease the excitability which associated with membrane of sodium ions across the membrane. Therefore local anesthetic interferes with sodium movements and interferes with excitability of cells.

6. The local anesthetic activity is dependant on its entering the channel from inside the neuron.

UNIT – II

CENTRAL NERVOUS SYSTEM-I

- General Anaesthetics
- Hypnotics and Sedatives
- Opioid Analgesics
- Nonsteroidal anti- inflammatory agents

CENTRAL NERVOUS SYSTEM-I

General Anesthetics

The term Anesthesia means loss of sensation. General anesthetic is a class of CNS depressant drugs which produce total loss of sense of pain with controlled and reversible loss of consciousness.

General anesthetics bring about descending depression of the CNS starting with cerebral cortex, basal ganglia, cerebellum and finally the spinal cord. General anesthetics are mainly used in surgical operations for relaxation of muscles of patients. These agents are non specific with receptors and used at high concentration to access to all areas of the body.

Classification

Based on the method of administration, these may divided in to three groups.

1. Inhalation Anesthetics

    (i)   Hydro Carbons – Cyclo propane, Ethylene

    (ii) Halogenated Hydro carbon – Halothane, Chloroform, Ethyl chloride

    (iii) Ethers – Di ethyl ether, Vinyl ether, Enflurane, Methoxy flurane, Iso flurane, Desflurane, Sevoflurane.

    (iv) Alcohols – Trichloro ethanol

    (v) Miscellaneous – Nitrous oxide

2. Intravenous anesthetics

   (i) Ultra short acting Barbiturates – Metho hexitol sodium, Thio amylol sodium, Thio pentol sodium

   (ii) Benzo diazepines – diazepam, Midazolam

   (iii) Miscellaneous – Ketamine, Etomidate, Propofol, Alphaxalone

3. Basal Anesthetics

Fentanyl citrate, Tribromo ethanol, paraldehyde

Inhalation Anesthetics

These are volatile liquid or Gases and they are administered through inhalation process.

1. Di ethyl ether or Anesthetic ether

$C_2H_5 - O - C_2H_5$

Synthesis

$C_2H_5\ OH + H_2SO_4\ C_2H_5\ H_2SO_4\ C_2H_5 - O - C_2H_5$

$C_2H_5\ ONa + C_2H_5Br\ C_2H_5 - O - C_2H_5$

2. Vinyl Ether

$CH_2\ CH - O - CH\ CH_2$

Synthesis

$Cl\ CH_2\ CH_2OH\ Cl$ ---- $CH_2\ CH_2O\ CH_2\ CH_2\ Cl$ ---- $CH_2 = CH– O - CH = CH_2$

3. Halothane

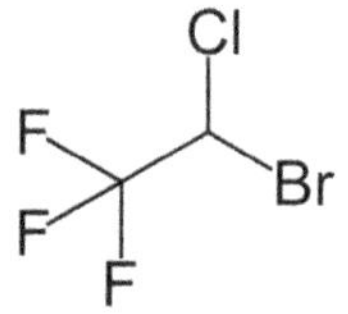

Halothane vapour (or **Fluothane**) is an inhalational general anaesthetic. Its IUPAC name is 2-bromo-2-chloro-1,1,1-trifluoroethane. It is the only inhalational anaesthetic agent containing a bromine atom; there are several other halogenated anesthesia agents which lack the bromine atom and do contain the fluorine and chlorine atoms present in halothane. It is colourless and pleasant-smelling, but unstable in light. It is packaged in dark-coloured bottles and contains 0.01% thymol as a stabilising agent. Halothane is a *core* medicine in the World Health Organization's "Essential Drugs List", which is a list of minimum medical needs for a basic health care system. Its use in developed countries however has almost entirely been superseded by newer inhalational anaesthetic agents.

4. **Nitrous oxide** (Laughing Gas) N_2O

5. **Cyclo propane**

   CH_2

   $CH_2\ CH_2$

6. **Tri chloro ethylene----- $CCl_2\ CHCl$**

II.   Intravenous Anesthetics

These are administered through intravenously and cause unconsciousness. These are mainly sodium salt of barbiturates. They are ultra short acting barbiturates which duration of action is less than 30 minutes.

1. **Thio pental sodium**

A. *Thiopental Sodium* INN, USAN, *Thiopentone Sodium* BAN,

Sodium 5-ethyl-5-(1-methylbutyl)-2-thiobarbiturate ,

Sodium thiopental, better known as **Sodium Pentathol** (a trademark of Abbott Laboratories), **thiopental, thiopentone sodium**, or **trapanal**, is a rapid-onset short-acting barbiturate general anaesthetic. Sodium thiopental is a depressant and is sometimes used during interrogations—not to cause pain (in fact, it may have just the opposite effect), but to weaken the resolve of the subject and make him or her more compliant to pressure.

2.   Metho Hexitol Sodium

$CH_3CH_2C\equiv C-MgBr$
1-Butynyl magnesium bromide

$\xrightarrow[\text{(ii) } PCl_5]{\text{(i) } CH_3CHO}$

$CH_3C\equiv C-CH-CH_3$ (with CH branch)
2-Chloro-3-pentyne

$NC-CH_2-\overset{O}{\underset{\|}{C}}-OC_2H_5$
Ethyl cyanoacetate

$+ C_2H_5ONa$
Sodium ethylate

Ethyl -(1-methyl-2-pentynyl)-cyanoacetate

$H_2C\equiv CH-CH_2-Br$
Allyl Bromide

Ethyl-(1-methyl-2-pentynyl)-allyl cyanoacetate

$O=C\underset{NH}{\overset{NH_2}{<}}$
CH_3
N-Methyl urea

Methadol

$\xrightarrow{NaCl}$

Metho hexitol sodium

Ketamine Hydro Chloride

Ketamine is a drug used in human and veterinary medicine developed by Parke-Davis (today a part of Pfizer) in 1962. Its hydrochloride salt is sold as **Ketanest, Ketaset,** and **Ketalar**. Pharmacologically, ketamine is classified as an NMDA receptor antagonist.[2] At high, fully anesthetic level doses, ketamine has also been found to bind to opioid μ receptors and sigma receptors.[3][4] Like other drugs of this class such as tiletamine and phencyclidine (PCP), it induces a state referred to as "dissociative anesthesia"[5] and is used as a recreational drug.

Synthesis

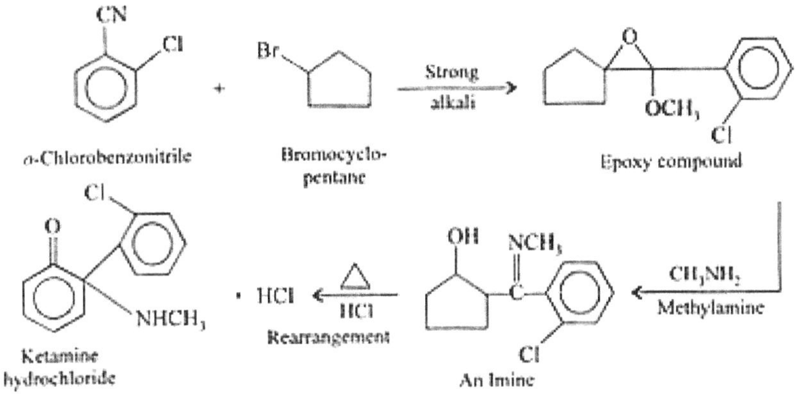

Mechanism of Action

General anesthetics are non specific in action because they do not act on specific receptor site.

When General anesthetics are dissolved in lipid layer and membranes, they cause disordering or increased fluidity of the membrane. This may inactive the protein essential functioning of CNS or ion channel, which would remain open resulting in chloride ion influx leading to synoptic hyper polarization, which inhibits neuronal function. Volatile General anesthetics are rapidly evaporates and cooled immediately and skin gets anesthetized.

Barbiturates mainly depress the CNS and decrease specific functional activities in brain. They increase the GABAnergic inhibitory response by influencing conductance at the chloride channel. They are also uncoupling of the oxidative phosphorylation, prevent the electron transport system and inhibit the cerebral carbonic anhydrase activity.

Anxiolytics, Sedatives and Hypnotics

These drugs are under the category of CNS depressants.

Sedatives are drugs which decrease activity and excitement of the patients and calm the anxiety by producing mild depression of CNS without causing drowsiness or sleep.

Hypnotics are drugs which produce drowsiness and induce sleep resembling to natural sleep by depressing CNS.

Anxiolytics are also called as minor tranquillizers which are used in treatment of psychotic disorder by depressing CNS. They are used to treat abnormalities of mental functions.

The various sedatives and hypnotic agents are also employed as:

   (i)   Antianxiety agents in emotional strain and chronic tension

  (ii)   Anticonvulsant

 (iii)   Muscle relaxants

 (iv)   General anaesthetics

  (v)   Potentiation of analgesic drugs

 (vi)   Adjuvants to anesthesia

(vii)   In hyper tension

Classification

I. Barbiturates

1. Long acting barbiturates (More than 8 hours) – Barbitone, Barbital sodium, Phorobarbital, Mephobarbital

2. Intermediate acting barbiturates (4–8 hours) – Allo barbital, hexobarbitone,
   pento hexital sodium

3. Short acting barbiturates (less than 4 hours) – Seco barbitone, hexobarbitone,
   Pento barbitone, cyclo barbitone

4. Ultra short acting barbiturates (I-V anaesthetics) – Thiopental sodium. Metho hexital sodium

II. Benzodiazepines

Diazepam, Chlordiazepoxide, oxazepam, chlorazepate dipotassium, prazepam, Lorazepam, Halazepam, Flurazepam, Alprazolam, Triazolam

III. Amides and Imides

Glutethimide, methyprylon, methaqualone

IV. Aldehydes and their derivatives

Chloral hydrate, Paraldehyde, Triclofos sodium

V. Alcohol and their carbamate derivatives

Ethinamate, Meprobamate

VI. Miscellaneous

Antihistamines, KBr, Morphine, Pethidine

I. Barbiturates

Barbiturates are cyclic diurides which are the derivative of barbituric acid (2,4,6 tri oxo hexahydro pyrimidine). As such barbituric acid has no CNS depressant activities. But substitution at 5th position by alkyl or aryl groups confers sedative and hypnotic activity.

Barbituric acid or barbiturates are prepared from malonic acid or its esters.

(a)   By the interaction of urea and malonyl dichloride

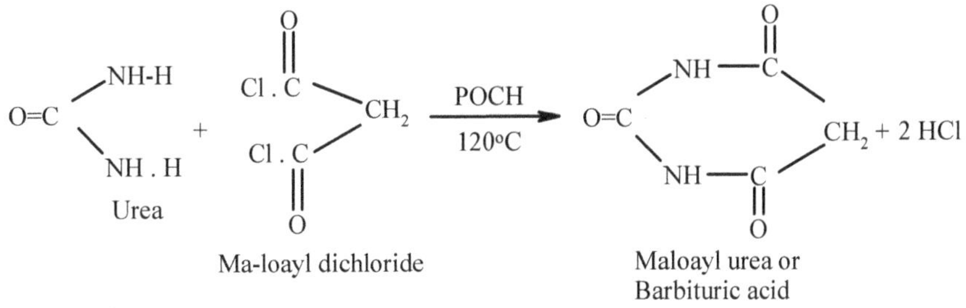

Ma-loayl dichloride   Maloayl urea or
                             Barbituric acid

(b)   By the interaction of urea and diethyl malonate

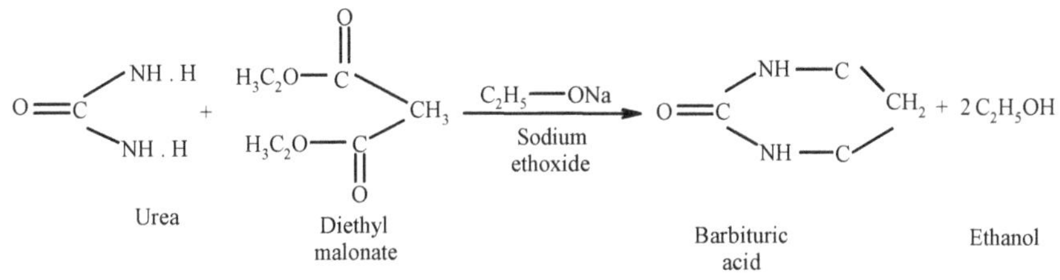

Urea   Diethyl malonate   Barbituric acid   Ethanol

The cyclic ureides containing a six membered ring, are also regarded as derivatives of the ion mental type pyrimidine or 1 : 3-diazine

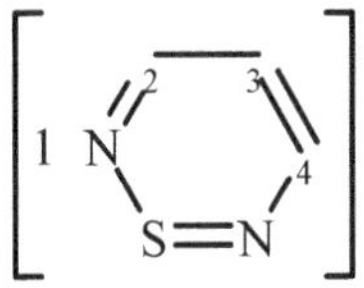

Barbituric acid like parabanic acid exhibits "zero-enol taumerism" as illustrated below:

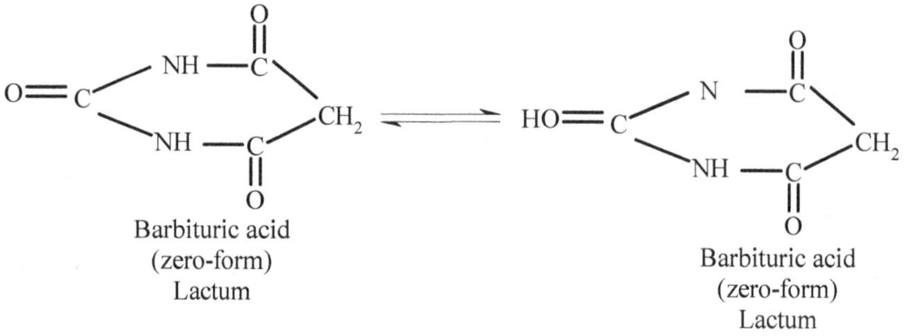

Barbituric acid
(zero-form)
Lactum

Barbituric acid
(zero-form)
Lactum

Phenobarbital

Phenobarbital or phenobarbitone is a barbiturate, first marketed as Luminal by Friedr. Bayer et comp. It is the most widely used anticonvulsant worldwide and the oldest still commonly used. It also has sedative and hypnotic properties but, as with other barbiturates, has been superseded by the benzodiazepines for these indications. The World Health Organization recommends its use as first-line for partial and generalized tonic-clonic seizures (those formerly known as Grand Mal) in developing countries. **IUPAC NAME : 5-ethyl-5-phenylpyrimidine-2,4,6(1*H*,3*H*,5*H*)-trione**

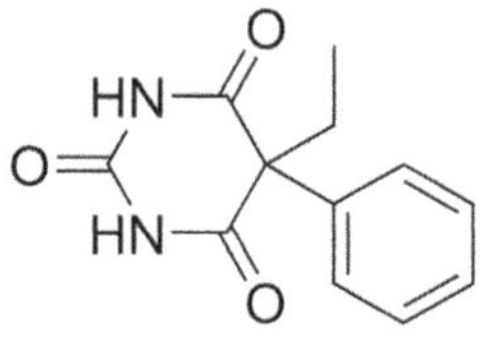

Synthesis of Phenobarbitone

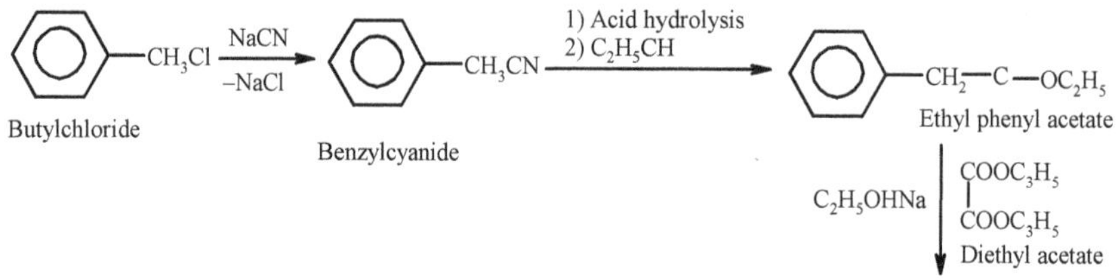

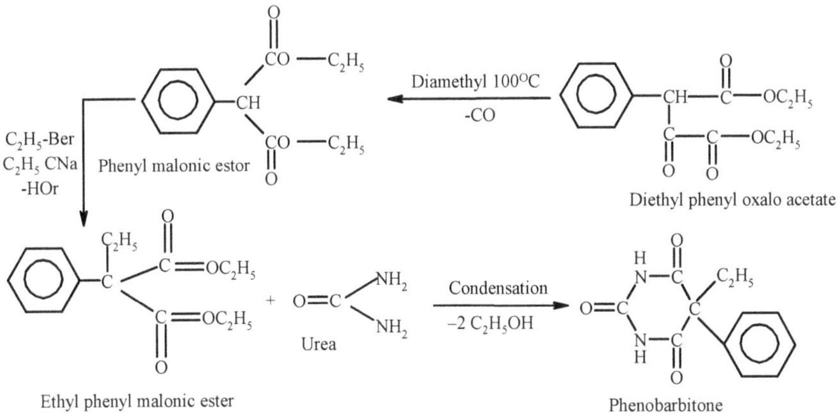

Diamethyl 100°C
-CO

Diethyl phenyl oxalo acetate

C_2H_5-Ber
C_2H_5 CNa
-HOr

Phenyl malonic estor

Ethyl phenyl malonic ester

Urea

Condensation
$-2 C_2H_5OH$

Phenobarbitone

E. Butabarbitone (Butabarbital)

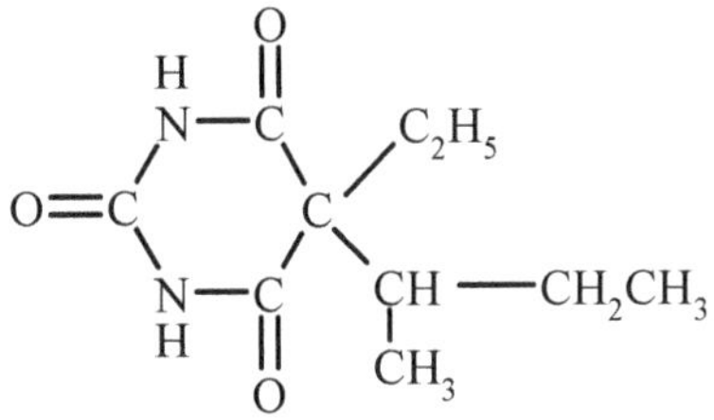

5(Methyl propyl(1-5- ethyl barbuturic acid

Synthesis

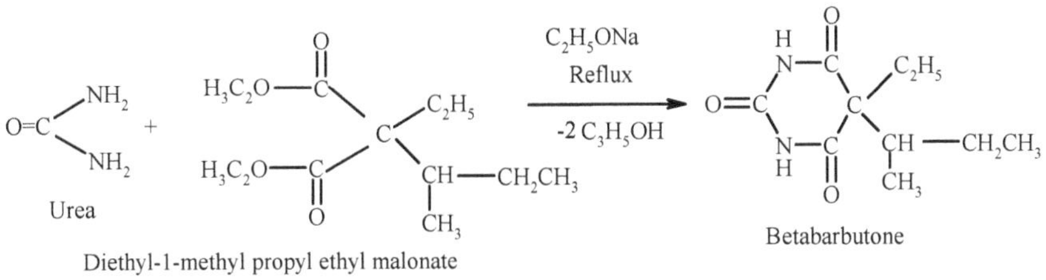

Urea

Diethyl-1-methyl propyl ethyl malonate

C_2H_5ONa
Reflux
$-2 C_3H_5OH$

Betabarbutone

Use : Same as Allobarbitone

Structure-Activity Relationship of Barbiturates (SAR)

1. Barbituric acid itself does not possess any sedative and hypnotic activity. The sedative or hypnotic activity is produced when the

two active hydrogen atoms at position 5,5 have appropriate substituents by alkyl or aryl groups.

2. The total number of carbon atoms present in two groups at 5th position must not be less than 4 and more than 10 for their optiomal activity.

3. Only one of the substituents may be a closed chain for good activity.
    (Example : Hexobarbital, Cyclobarbital, Phenobarbital).

4. The branched chain isomer exhibits greater activity and shorter duration. The drug having greater branching is more potent (Example : Pento barbital, amobarbital).

5. Double bonds are present in alkyl substituents groups produce compounds more readily to tissue oxidation. Hence they are short acting (Example : Seco barbital)

6. Stereo isomers have appropriately same potencies.

7. Aromatic or alicyclic substituted analogues are more potent than the corresponding aliphatic analogues having same number of carbon atoms.

8. Short chain at carbon 5 resists oxidation and hence long acting. Long chains are readily oxidized and thus produce short acting (Example : Barbital, Pentobarbital, Seco barbital)

9. Introduction of halogen atom is to 5 – alkyl substituents increases potency

10. Introduction of polar groups (OH, NH, COOH, SO_3H, CO, RNH) is to 5 – alkyl or aryl substituents decrease lipid solubility and Potency.

11. Alkylation at 1 or 3 position enhances onset and reduce duration of action
    (Example : Hexobarbital).

12. Replacement of oxygen by sulphur at 2 – carbon (thio barbiturates) shortens on set and duration of action due to increased lipid solubility. But more sulphur at C4SC6 decrease the activity (Example: Thio pental)

Mechanism of Action

1. Barbiturates either have a Gama amino butyric acid (GABA) like action or enhance the effect of GABA, an inhibitory transmitter.

2. The effects of barbiturates on synaptic transmission are caused by an alteration of post synaptic sensitivity of the neurons to excitatory and inhibitory transmitters.

3. When GABA receptors are activated, chloride channels are open and chloride enters the cell, hyper polarizes it and produces decreased excitation.

4. Barbiturates bind to picrotoxin site of GABA receptor and decrease chloride ion flux and produce an increased chloride ion concentration.
   They may interfere with the passage of impulses from centers in hypothalamus to the cortex.

Narcotic Analgesic (Morphine and Related Drugs)

Analgesics are agents which relieve the pain without disturbing consciousness.

Analgesics are divided in to two main classes

1. Narcotic analgesics (Centrally acting drugs)

2. Non Narcotic analgesics (Peripherally acting drugs)

The narcotic analgesics are also called as opiate analgesics. These are mainly obtained from unripe capsules of papaver somniferum (Opium poppy) plant. The important alkaloid is isolated from opium is morphine. The other alkaloids isolated from opium are codeine, Papaverine and thebain.

The opium group of narcotic drugs is most powerfully acting and clinically useful drugs producing depression of CNS. They depress the CNS and relieve the pain and some drugs like morphine induce sleep in presence of pain, diarrhea and suppress sough. The term opiod is used generally to designate collectively the drugs (natural or synthetic) which bind specifically to any of sub species of receptor of morphine and produce morphine like actions.

The limitations of opiate analgesics are -they have addictive property, respiratory depression, decreased gastro intestinal motility leading to constipation, increases biliary tract pressure and pruritis due to histamine release.

Classification

1. Morphine Analogues

Morphine SO_4, Codiene PO_4, Ethyl Morphine, Diacetyl morphine(Heroin), Hydro morphoneHCl, Oxy morphone.HCl, Apo morphine.HCl, Hydrocodone, Oxy codone, Dihydromorphine, Dihydro codeine.

2. Morphinan Analogues

Levorphanol tartarate, Dextro methorphan, Butarphanol,

3. Morphan Analogues

Metazocin, Cyclazocin, Pentazocin.

4. 4-Phenyl Piperidine Analogues

Meperidine.HCl(Pethidine.HCl), Di phenoxylate.HCl, Fentanyl citrate, Anileridine.HCl, Phenoperidine, Alphaprodine.HCl, Loperamide.HCl.

5. Phenyl propylamine Analogues

Methadone.HCl, Dextro propoxyphene.HCl, Metho Trimeprazine.

6. Miscelleneous:

Tramadol, Tilidate, Nexeridine, Sulfentanil.

7. Narcotic Antagonists

Nalorphine, Naloxone, Levellorphan, Naltrexene, Cyclazocine, Propiram, Profadol

I. Morphine Analogues:

Morphine

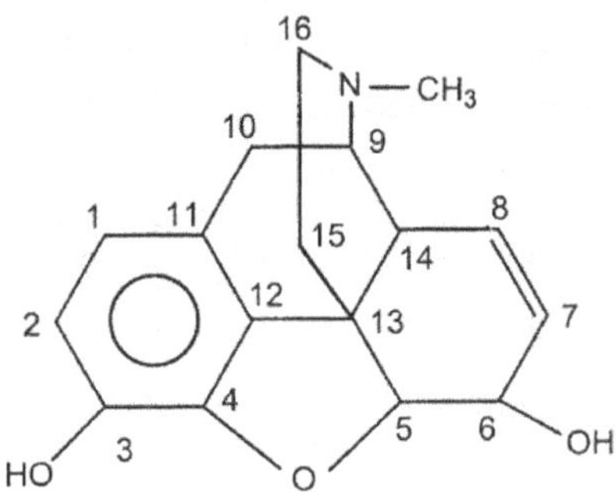

Derivatives of morphine

S.No	Compound Name	R_3	R_6
1.	Codeine	CH_3	H
2.	Ethylmorphine	C_2H_5	H
3.	Heroin	CH_3CO-	CH_3CO-

S.No	Compound name	R_3	R_5	R_{14}
1.	Hydromophone	H	H	H
2.	Hydrocodone	CH_3	H	H
3.	Oxycodone	CH_3	H	OH
4.	Oxymorphone	H	H	OH
5.	Methyl dihydromorphinone	H	OCH_3	H

II. Morphinan Analogues

Levorphanol Tartarate

17 - Methyl morphinan -3-ol tartrate dihydrate

III. Morphan Analogues

Metazocine

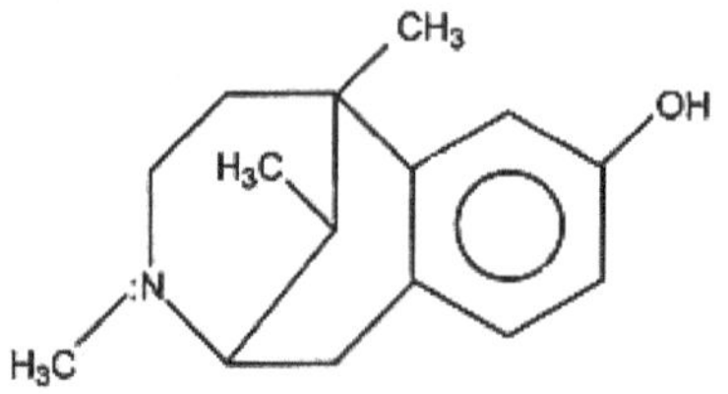

2' - Hydroxy - 2,5, 9 - trimethyl - 6,7 - benzomorphan

Pentazocine

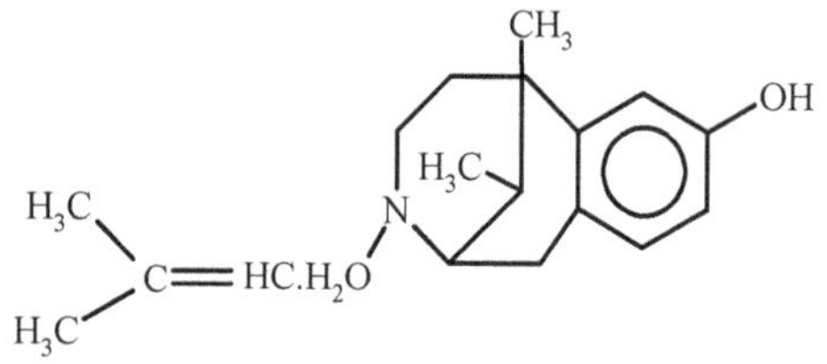

1,2,3,4,5,6 - Hexahydro-6, 11 -dimethyl - 3-(3-methyl-2-butenyl) -2-6-methano-3-benzazocin - 8 - ol

Pentazocine is a synthetically-prepared prototypical mixed agonist-antagonist narcotic (opioid analgesic) drug of the benzomorphan class of opioids used to treat mild to moderately severe pain. Pentazocine is sold under several brand names, such as **Fortral, Talwin NX** (with the mu-antagonist naloxone, will cause withdrawal in opioid dependent persons), **Talwin, Talwin PX** (without naloxone), **Fortwin** (Lactate injectable form) and **Talacen** (with acetaminophen). This compound may exist as one of two enantiomers, named (+)-pentazocine and (-)-pentazocine. (-)-pentazocine is a kappa-opioid receptor agonist, while (+)-pentazocine is not, instead displaying a ten-fold greater affinity for the sigma receptor. Talwin PX is the main pentazocine pharmaceutical in Canada, where laws and regulations prohibit the addition of naloxone to the formulation for non-therapeutic purposes. Related drugs include phenazocine, dezocine, cyclazocine, salvinorin A (distantly) and several chemicals used in research on the central nervous system, and the Greek equivalent of the first letter of the name of the drug ketocyclazocine is the source of the name of the kappa opioid receptor type, as is the case with morphine and mu receptors and dynorphin and delta opioid receptors

IV.   4-Phenyl Piperidine Analogues

Loperamide

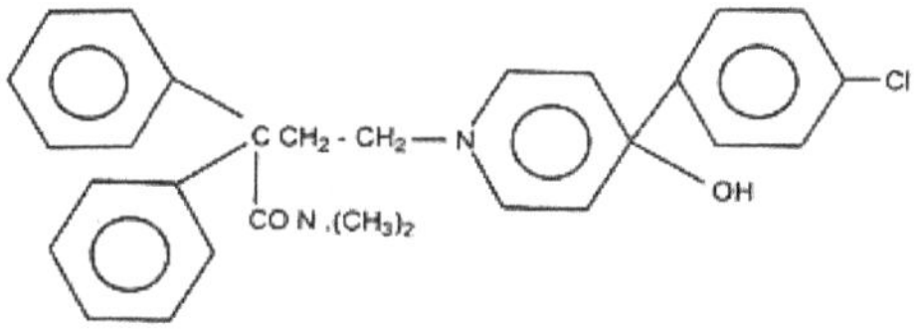

4-(4-Chlorophenyl)-4-hydroxy-N,   N-dimethyl   – α, α-diphenyl-1-piperidine butanamide

A. Meperidine Hydrochloride (Pethidine hydrochloride)

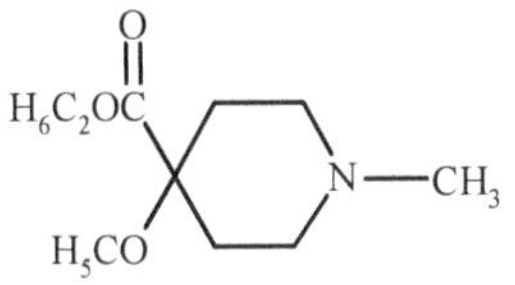

4 - Ethoxy carbonyl-1-methyl -4- phenyl piperidine

Synthesis

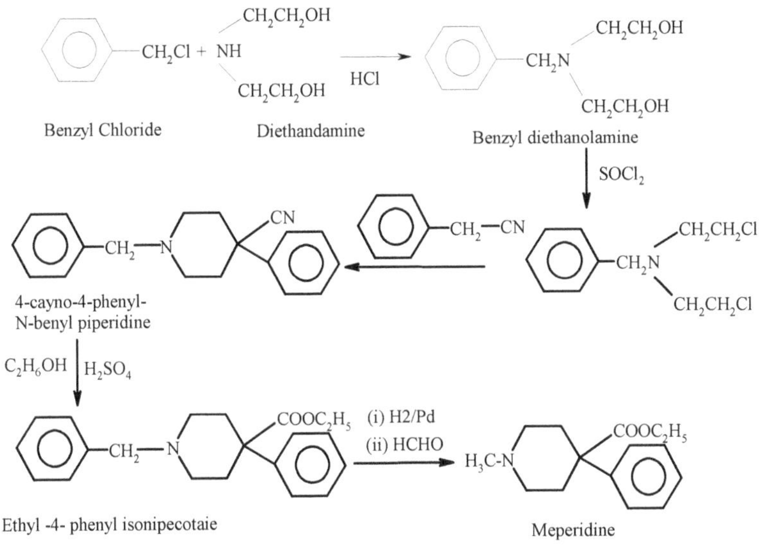

Benzyl Chloride   Diethandamine   Benzyl diethanolamine

4-cayno-4-phenyl-
N-benyl piperidine

Ethyl -4- phenyl isonipecotaie

Meperidine

V.   Phenyl propylamine Analogues:
Methadone

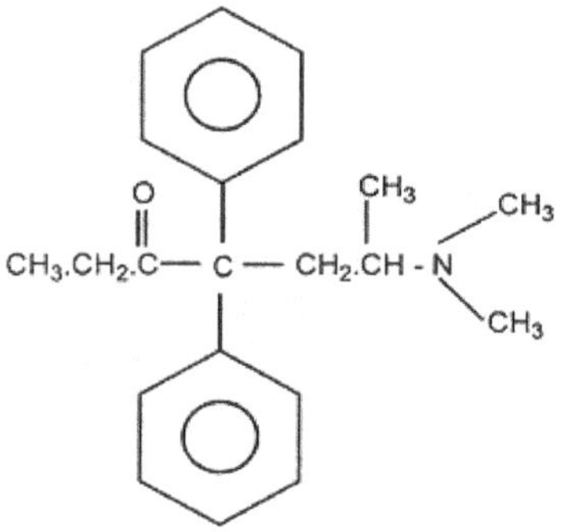

$$CH_3.CH_2.C - C - CH_2.CH - N$$

6-Dimethylamino -4,4 -diphenyl heptan -3-one

VI.   Narcotic Antagonist

Nalorphine Hydrochloride

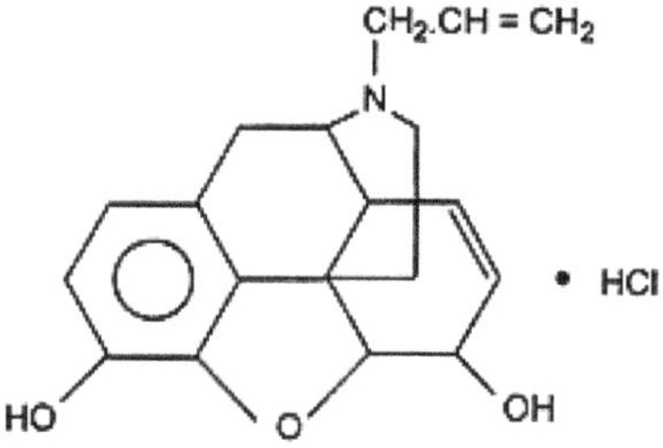

17 - Allyl -7,8 - didehydro - 4,5 - epoxymorphinan - 3,6 -diol hydrochloride

Naloxone Hydrochloride

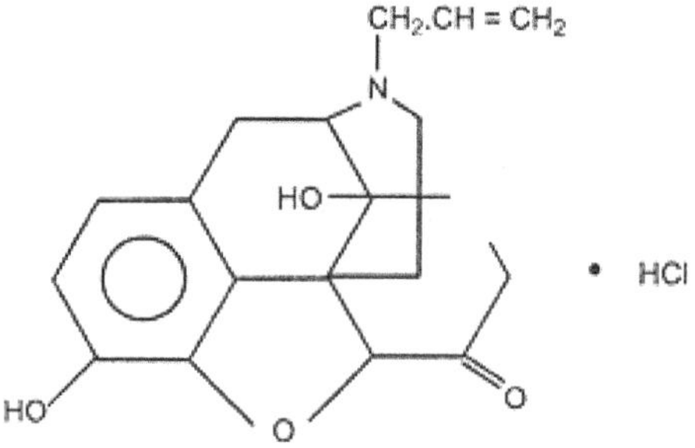

17 - Allyl -5-epoxy - 3,14 - dihydroxy morphinan - 6 - one hydrochloride

SAR for Morphine like drugs:

General Structure

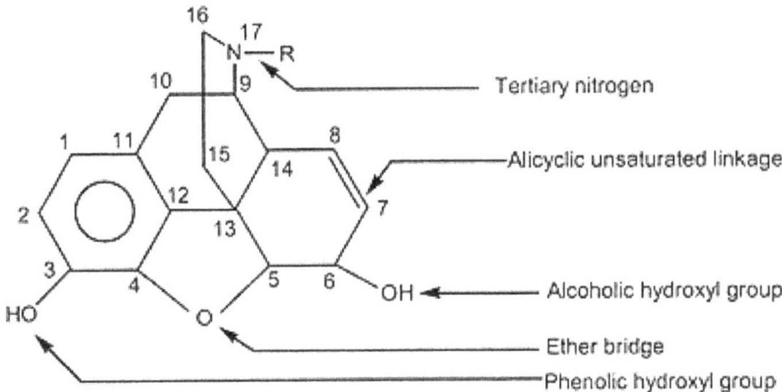

The structural activity relationship is studied due to the modifications of the following parts of morphine.

1. Modifications on aromatic ring system

2. Modifications on alicyclic ring system

3. Modifications of Tertiary nitrogen

4. Modifications of Ether Bridge

I. Modifications on aromatic ring system

1. An aromatic phenyl ring is essential for activity.

2. Modifications of C3 phenolic hydroxyl group decreases analgesic activity.

3. Making the phenolic – OH group by etherification to methyl ether (Codeine) and ethyl ether (ethyl morphine) results in about one tenth of analgesic activity of morphine. Because phenolics – OH group binds with opiate receptor by hydrogen bonding easily. But ethers are not easily hydrolysis.

4. Esterification of 3 – OH group gives compounds more active than morphine.

5. Substances other than 3-position in the aromatic ring results in a reduction of opioid actions. But 1-fluoro codeine possesses the some analgesic activity as that codeine.

II. Modifications on alicyclic ring system

1. The C-6-Alpha- OH group is methylated, esterified, oxidized, removed or replaced by halogen in order to get more potent analgesics. But there is also a parallel increase in toxicity. Example: Codeine, heroin, chloro morphone.

2. The saturation of double bond at C – 8 position gives more potent compounds. Example: Dihydromorphine, Dihydrocodeine.

3. Introduction of 14 – OH in dihydro from gives more potent 14 – hydroxy dihydro codeinone and 14 – hydroxy dihydro morphinone.

4. Bridging of C6 and C14 through ethylene linkage gives etorphine which is 200 times more potent than morphine.

5. Introduction of any new substituents at 5th position does not enhance the activity except 5 – methyl dihydro morphine and azidomorphines.

III. Modifications of 30 Nitrogens

1. Replacement of $N-CH_3$ by $N-C_2H_5$ results slight fall in analgesic response. More hydrophobic groups such as propyl, pentyl, hexyl and phenylethyl gave an increase in activity.

2. N-allyl and N-cycloalkyl methyl functions give the narcotic antagonistic properties.

3. N-Phenyl ethyl group enhances the analgesic activity in desmorphine, codeine and heterocodeine.

IV. Modifications of ether bridge

1. Breaking of ether bridge and opening of piperidine ring decreases the activity.

SAR for meperidine analogues

1. Replacement of 4-phenyl group by hydrogen, alkyl, aralkyl or heterocyclic groups reduces the activity.

2. The presence of phenyl and ester group at position 4 of 1-methyl piperidine gives optimum activity.

3. The replacement of N-methyl group by various aryl alkyl groups can increase the analgesic property.

4. Introduction of m –hydroxy group in phenyl ring increases the activity similar to C_3 – OH of morphine. Example : Bemidone.

5. Replacement of ester moiety by a ketone in bemidone. Example: ketobemidone is equalent to morphine in activity.

6. The reversed ester of meperidine, propionoxy compound was more active, being 5 times more active than meperidine. Example: Prodine.

7. When phenyl and acyl groups are separated from piperidine ring by a nitrogen atom, it gives a powerful analgesic. Example: Fertanyl.

8. By enlarging the piperidine ring to seven member hexahydro azepine ring. Example: Proteptazine is more active analgesic agent.

9. Contraction of piperidine ring to five member Pyrrolidine ring was also have good activity. Example: Alpha prodine, Prodilidine

SAR for methadone series

1. The Levo isomer of methadone and Isomethadone are twice active as its racemates.

2. Removal of any one of phenyl rings decreases the activity.

3. Introduction of m – hydroxy group in phenyl ring decreases the activity.

4. Methadone derivatives are generally more potent analgesic than isomethadone series.

5. The replacement of propionyl group by hydrogen, hydroxy or acetyloxy group leads to decrease the analgesic activity.

6. Replacement of propionyl group by amide group (ex. Racemoramide) is more active than methadone.

7. Replacement of dimethylamino group by heterocyclic ring like morpholine and piperidine are potent as methadone with morphine like activity (Racemoramide).

8. An N-methylated derivative of metabolites of methadone analogues retains the analgesic activity.

Mechanism of Action of Opiods

1. The Pharmacological actions of opiods are mediated by several types of opiate receptors in the CNS.

2. The structural features which are recognized to be essential for the perfect fit of a narcotic analgesic on receptors are represented below.

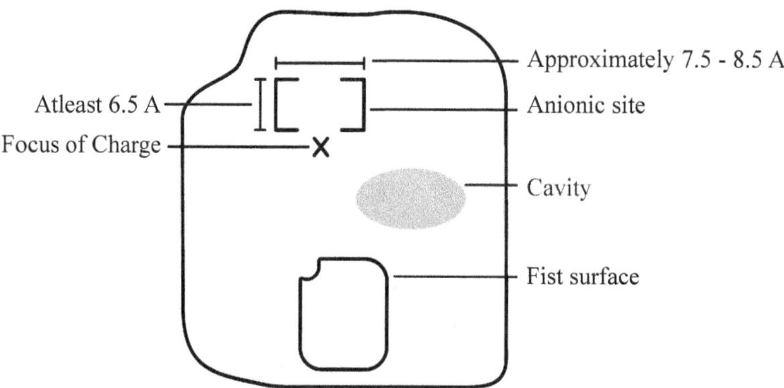

Where   A = Phenyl or aromatic portion

B = Quaternary carbon

C = Ethylene bridge

D = Tertiary nitrogen

Beckett and Casy (1954) proposed that an opioid receptor is composed of three prominent parts.

3. Opioid receptors compared of three major arts. i) A flat portion with holds aromatic part by Vandar-Waal's force. ii) A cavity or a hallow portion with entraps ethylene bridge. iii) An ionic site which holds the 30 nitrogen with get ionized at physiological. pH Beckett and Casy model of the analgesic receptor site is in above figure.

4. The fact that these sites do not bind other substances and are saturated by even very low concentrations of opiods explains the highly stereospecific orientation of these three components of opiod receptors.

5. There are three major types of opiod receptors (i) Mu (m) – op3 receptors – produce analgesia, respiratory depression, Euphoria and addiction.ii) kappa (K) – op2 receptors – produce dysphoria, Euphoria and addiction.iii) Delta (d) – op1 receptors – G – proteins – linked receptors.

6. Morphine binds to m receptor and induces change in shape and open the ion channel in cell membrane. So K+ ion can flow out of the cell, hyperpolarizes membrane potential. Therefore the frequency of action potential firing is decreased; result in a decrease in ion neuron excitability.

7. The increase in permeability decreases the influx of Ca into nerve retinal and reduces neuro transmitter release. Both the effects shut down the nerve and block pain message.

8. Kappa receptor is directly associated with Ca channel. When an agonist binds to K receptors, the Ca channel is closed. Since Ca is necessary for neurotransmitter it cannot pass on pain message.

9. When agonist binds with d(delta) receptors, the receptor changes its shape and triggers a messenger protein (G protein) to carry a message to a neighboring enzyme with catalyses the formation of cyclic adenosine monophosphate. The G protein inactivates the enzyme by preventing the synthesis of cyclic AMP. This act as a second messenger is the transmission of pain signed and stops the pain.

Anti Inflammatory Agents

Inflammation may be defined as the series of changes that occur in living tissues following injury.

The injury which is responsible for inflammation may be variety of conditions such as

1. Physical agents like UV radiation, heat, mechanical trauma.

2. Chemical agents like organic and inorganic compounds.

3. Toxins of various bacteria, intracellular replication of virus.

In old days some steroids like prednisolone, dexamethasone, betamethasone, triamcinalone and hydrocartisone were used as anti inflammatory agents. But these drugs produce some adverse effects. But nowadays much safer and better tolerated Non Steroidal Anti Inflammatory Drugs (NSAID) are used.

The non narcotic analgesics are having the following groups.

1. Analgesics which relieve the pain without interacting the opioid receptors.

2. NSAID which possess anti inflammatory property.

3. Anti pyretic which reduce the elevated body temperature.

Classification

1. Salicylic acid Derivatives

Sodium salicylate, Sodium thio salicylate, Magnesium salicylate, Choline salicylate, Carb ethyl salicylate, Phenyl salicylate, Salicylamide, Aspirin, Aluminium aspirin, Calcium acetyl salicylate, Salsalate.

2. N-Anthranilic acid Derivatives

Mefenamic acid, Meclofenamate sodium

3. Aryl acetic acid Derivatives

Indomethacin, Sulindac, Tolmetin sodium, Zomepriac sodium, Ibuprofen, Naproxan, Fenoprofen calcium, Ketoprofen, Flurbiprofen, Diclofenac sodium and potassium, Ketorolac tromethamin, Piroxicam.

4. Aniline and Para Amino Phenol Derivatives

Acetanilide, Phenacetin, Acetaminophen(Paracetamol).

5. Pyrazolone and Pyrazolidine dione Derivatives

Antipyrin, Aminopyrin, Dipyrone, Phenyl Butazone, Oxyphen Butazone.

I. Salicylic acid Derivatives

Aspirin, also known as **acetylsalicylic acid**, is a salicylate drug, often used as an analgesic to relieve minor aches and pains, as an antipyretic to reduce fever, and as an anti-inflammatory medication. Aspirin also has an antiplatelet effect by inhibiting the production of thromboxane, which under normal circumstances binds platelet molecules together to create a patch over damage of the walls within blood vessels. Because the platelet patch can become too large and also block blood flow, locally and downstream, aspirin is also used long-term, at low doses, to help prevent heart attacks, strokes, and blood clot formation in people at high risk for developing blood clots.[1] It has also been established that low doses of aspirin may be given immediately after a heart attack to reduce the risk of another heart attack or of the death of cardiac tissue

Aspirin (Acetyl salicylic acid)

Acetyl salicylic acid

It is a prodrug of salicylic acid

Synthesis

Salicylic acid → (Acetylation $(CH_3 CO)_2O$, CH_3COOH) → Aspirin

SAR of Salicylic acid Derivatives

1. Various substitution on the carboxyl or hydroxyl group result in to change in potency as well as toxicity.

2. The hydroxy group in ortho position is very important for activity.

3. Salt of salicylic acid with choline and magnesium possess longer duration of action and lesser gastro intestinal irritation than aspirin.

4. Salsalate is an ester of two salicylic acid molecules. Since it is insoluble in stomach and is not absorbed until it reaches the small intestine. So it cause less gastric irritation.

5. Introduction of hydrophobic group(F) at 51 positionof flufenasil is more potent, longer acting and with less gastric irritation.

Mechanism of action

The antipyretic and anti inflammatory actions of salicylates and other acidic drugs are probably due to their inhibitory effect on prostaglandin synthesis by inhibiting prostaglandin synthetase enzyme.

Salicylates exert their antipyretic action by increasing heat elimination of the body through the mobilization of water and consequent dilution of the blood.

II. Aryl acetic acid Derivatives

Indomethacin

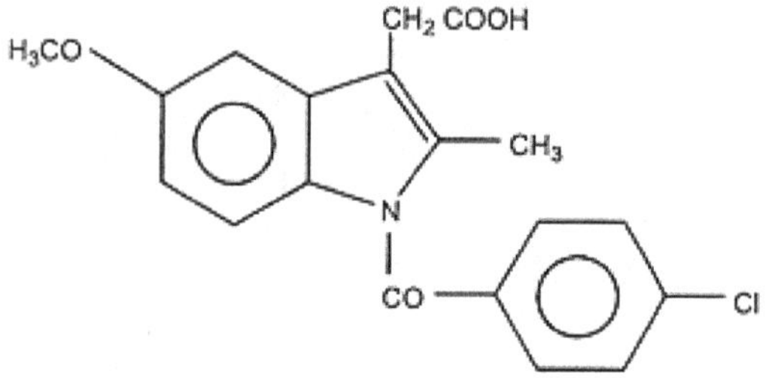

1 - (p- Chlorobenzoyl) - 5 - methoxy - 2 - methyl indole - 3 - acetic acid.

Ibuprofen

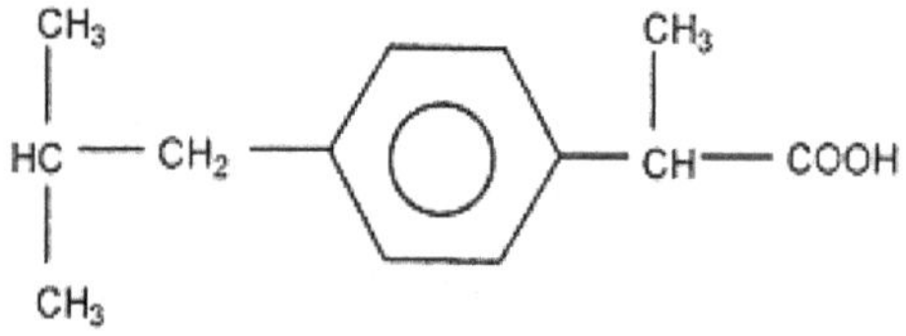

2 - (4- Isobutyl phenyl) - propionic acid

Synthesis

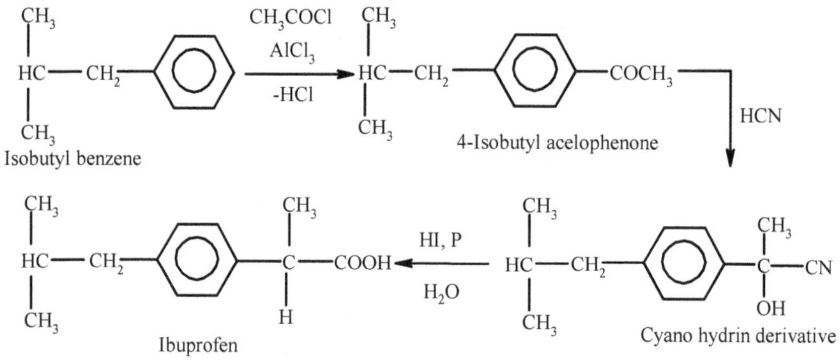

Isobutyl benzene

CH_3COCl / $AlCl_3$, $-HCl$

4-Isobutyl acelophenone

HCN

Cyano hydrin derivative

HI, P / H_2O

Ibuprofen

Ibuprofen ; from the now outdated nomenclature **iso-bu**tyl-**pro**panoic-**phen**olic acid) is a non-steroidal anti-inflammatory drug (NSAID) originally marketed as **Brufen,** and since then under various other trademarks (see tradenames section), most notably Nurofen, Advil and Motrin. It is used for relief of symptoms of arthritis, primary dysmenorrhea, fever, and as an analgesic, especially where there is an inflammatory component. Ibuprofen is known to have an antiplatelet effect, though it is relatively mild and short-lived when compared with that of aspirin or other better-known antiplatelet drugs. Ibuprofen is a *core* medicine in the World Health Organization's "Essential Drugs List", which is a list of minimum medical needs for a basic health care system

Ketoprofen

2-(3 – Benzoyl phenyl) propionic acid

Dichlofenac

O – (2,6, - Dichioro anilino) phenyl acetic acid

Diclofenac (marketed as **Voltaren** and under a number of other trade names, see below) is a non-steroidal anti-inflammatory drug (NSAID) taken to reduce inflammation and as an analgesic reducing pain in conditions such as arthritis or acute injury. It can also be used to reduce menstrual pain, dysmenorrhea. The name is derived from its chemical name: 2-(2,6-**dichlor**anilino)**phenylac**etic acid.

Synthesis of Diclofenac Sodium

SAR for Aryl acetic acid Derivatives

I. For Indole acetic acid derivatives

1. N-substitution of indole derivative increase Anti inflammatory activity in the order of benzyl > alkyl > H

2. The methyl group at 2nd position of indole increase the activity.

3. The substitution at 5th position increase the activity in the order of $OCH_3 > (CH_3)_2$ N $>CH_3 > H$.

4. The carboxyl group is necessory for good Anti inflammatory activity.

5. The N-Benzoyl group of indolehave halogen, CF3 or SCH3 at para position provides the greatest Anti inflammatory activity.

II. For Phenyl propionic acid derivatives

1. The maximum activity is obtained for the substitution at R1 is isobutyl group. The smaller substituents (Methyl, Ethyl) reduces the activity.

2. Maximal activity is found with R2 is methyl group. Smaller and larger groups diminish the activity.

3. Replacement of carboxyl group by an ester, alcoholic, amide, Hydroxamic acid(NHOH) or tetrazole(CHN_4) generally produce less active compounds.

4. The anti inflammatory activity resides in the S(+) isomer.

III. For Naphthyl propionic acid derivatives

1. The anti inflammatory activity is reduced when R1 is larger than OCH_3 or SCH_3 group.

2. The activity may be reduced when the carboxyl group is replaced with alcohol or aldehyde.

3. Dextro rotatory isomer is 11 times more active than phenyl butazone.

IV. For Oxicams

1. The nitrogen of benzothiazine ring have the substituent CH_3 and other electron withdrawing groups on the anilide phenyl groups such as Cl, CF_3, have good anti inflammatory activity.

2. The introduction of a heterocyclic ring in the amide oxide chain significantly increase the activity. (Sudoxicam is more potent than indomethacin).

3. The benzothiazine have pKa range of 6 to 8 have more activity.

VI. Aniline and P-Aminophenol Derivatives

Phenacetin

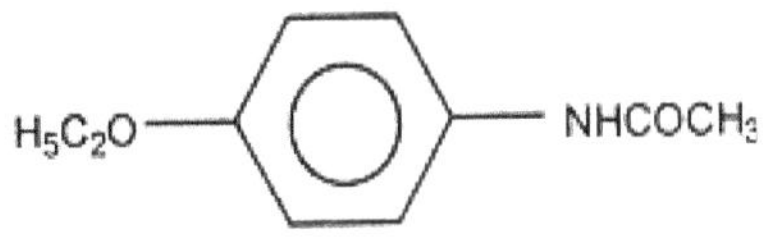

$$H_5C_2O-\langle\bigcirc\rangle-NHCOCH_3$$

p – Ethoxy acetanilide

A. Paracetamol (Acetaminophen)

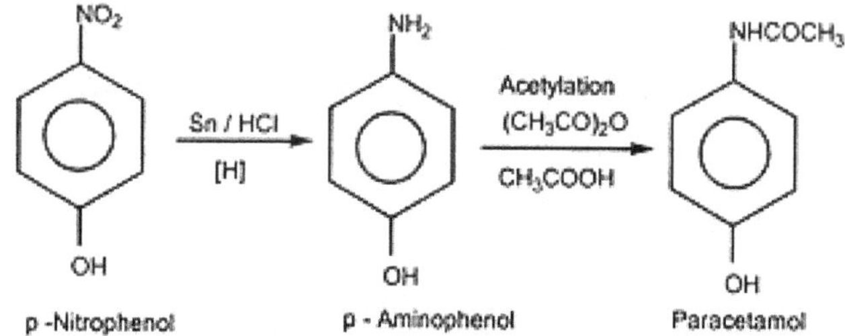

4- Hydroxy acetanilide

Synthesis

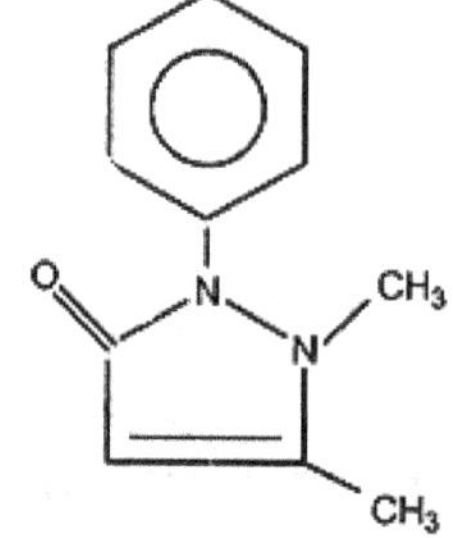

IV. Pyrazolone and Pyrazolidine dione Derivatives

Phenazone (Antipyrine)

2, 3 - Dimethyl - 1 - phenyl - 3 - pyrazolin - 5 - one

SAR for Pyrazolidine diones

1. The butyl group of carbon 4 may be replaced by propyl or allyl are less active.

2. The meta substitution of the aryl ring are inactive but para substitution such as CH_3, Cl, NO_2 or OH retains activity.

3. Replacement of nitrogen in pyrazolidines with oxygen yield isoxazole analog which is as active as pyrazolidine derivatives.

4. Decreasing pKa values of phenyl butazone analogs have shorter half lives.

5. Substitution at C-4 by methyl group destroys anti inflammatory activity.

6. The most active compound have the log P value is 0.7.

Mechanism of actions for NSAIDs

1. The prostaglandins are the mediation of inflammation. The NSAIDs are mainly prevent the formation of unstable proataglandin endoperoxide synthetase enzyme.

2. Most of the drugs are irreversible inhibitors of cyclo oxygenase activity, thus they prevent the formation of prostaglandins and consequently reducing the signs and symptoms of inflammation.

3. Besides inhibiting prostaglandin bio synthesis, the NSAIDs also inhibit the synthesis of leucotrienes, histamine and some biological process such as phagocytosis and platelet aggregation.

4. Other possible mechanisms of NSAIDs are kinin antagonism, prevention of leucocyte accumulation, stabilization of lysosome membranes, uncoupling of oxidative phosphorylation and oxygen radical scavenger action.

5. The inhibition of cyclo oxygenase can occur either by irreversible inactivation of enzyme (Eg-Aspirin) or rapid reversible non competitive inhibition involves anti oxidant or free radical trapping properties.

6. Some NSAIDs (Ibuprofen) reversible competitive inhibit by the propionic acid, which binds reversibly to the enzyme cyclo-oxygenase competing with arachidonic acid.

UNIT - III

CENTRAL NERVOUS SYSTEM-II

- Neuroleptics
- Antidepressants
- Antispasmodic and Antiulcer Drugs

CENERAL NERVOUS SYSTEM-II

CNS Stimulants and Psychedelics

The drugs that produce stimulation of central nervous system and enhancement in excitability of different portions of the brains or the spinal cord.

The CNS stimulants include analeptics, antidepressants, central sympathomimetic agents (Psychomotor stimulants). Some times CNS stimulants lead to convulsions so they are limited therapeutic value because of their convulsant activities and side effects.

Classification

1. *Analeptics* : Picrotoxin, Nikethamide, Etamivan, Pemoline, Pentylene tetrazole (Pentetrazole), Doxopram, Bemegride, Strychnine.

2. *Methyl xanthines* : Caffeine, Theophylline, Theobromine, Aminophylline, Etophylline, Proxyphylline.

3. *Central sympathomimetic agents (Psychomotor stimulants)* : Amphetamine, Methamphetamine, Phentermine, Benz phentamine, Chlorpentermine, Ferfluramine, Chlortermine, Phenmetrazine, Phendimetrazine, Mazindol, Methyl phenidate.

4. *Mono amino oxidase Inhibitors (MAO–inhibitors)* : Phenelzine, Isocarboxazid, Tranyl cypromine, pargyline, clorgyline.

5. *Tricyclic Antidepressants* : Imipramine, Desipramine, Amitryptyline, Nortriptyline, Protriptyline, Trimipramine, Doxepin, Maprotiline.

6. *Psychedelics* :

   1. *Indole ethyl amines* – Bufotenine, Psilocybin, Psilocyn

   2. *2 – Phenyl ethylamines* – Mescaline

   3. *Agents have both indolethylamine and phenyl ethylamine* – (+) Lysergic acid, Diethylamide (LSD).

   4. *Dissociative agents* – Phencyclidine (PCP)

   5. *Depressant-Intoxicant* – Tetra hydro cannabinol (THC)

Analeptics

Analeptics are agents which stimulate various areas of the central nervous system. These are mainly used for the treatment of respiratory depression resulting from overdose of depressant drugs. So these are used as respiratory stimulants.

An excessive dose of analeptics may result a wide–spread stimulation of the brain that may ultimately cause convulsions.

Mechanism of action for analeptics

1. Some drugs block post synaptic inhibition (Strychmine) or pre synaptic inhibition (Picrotoxin).

2. Some drugs acts as GABA ontagonist (Picrotoxins bemegride) or release prostoglandins and also decrease energy levels (rentetrazole)

Nikethamide

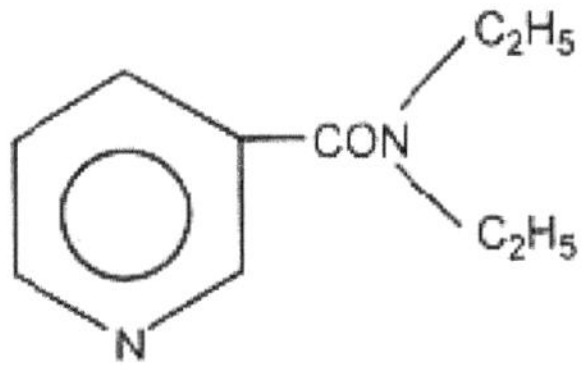

N, N - Diethyl nicotinamide

SAR for central Sympathomimetic Agents

Any decrease in distance between aromatic ring and heterocyclic nitrogen decrease the activity.

- The branched CH_3 group or similar substitution is important for activity, since it provides resistance to enzymatic is activation by stearic protection of the amino group.

- In phenidate series, activity is maximal at the methyl ester.

- In morpholine series aromatic substitutions and replacement of ring by heterocyclic groups decrease the activities.

Mechanism of action

1. They inhibit reuptake mechanisms for several biogenic amines.
2. They enhance neuronal release of catecholamines.
3. They stimulate a-adrenergic receptor and inhibit mono amino oxidase is   higher concentration.

Mono amino oxidase Inhibitors

Isocarboxazid

H$_3$C — O — N ... CO NH_ NH_ CH$_2$ —

5-Methyl - 3[2(phenyl methyl) hydrazide] isoxazole carboxylic acid

SAR for mono amino oxidase inhibitors

1. Cyclo alkyl substituents have equal potency with corresponding N-alkyl group Di alkyl substituted hydrazine (R2NNH2) devoid of significant activity.

2. Hydroxyl alkyl hydrazines are usually less effective MAO inhibitors than the corresponding alkyl hydrazines.

3. Aromatic ring substituents with polar groups decrease the activity Unsubstituted hydrazides ($RCONHNH_2$) do not inhibit MAO.

4. Mono substituted hydrazides may enhance MAO inhibitory activity.

Tricyclic Anti depressant

Imipramine

Imipramine (sold as Antideprin, Deprimin, Deprinol, Depsonil, Dynaprin, Eupramin, Imipramil, Irmin, Janimine, Melipramin, Surplix, Tofranil) is an antidepressant medication, a tricyclic antidepressant of the dibenzazepine group. Imipramine is mainly used in the treatment of major depression and enuresis (inability to control urination).

Imipramine

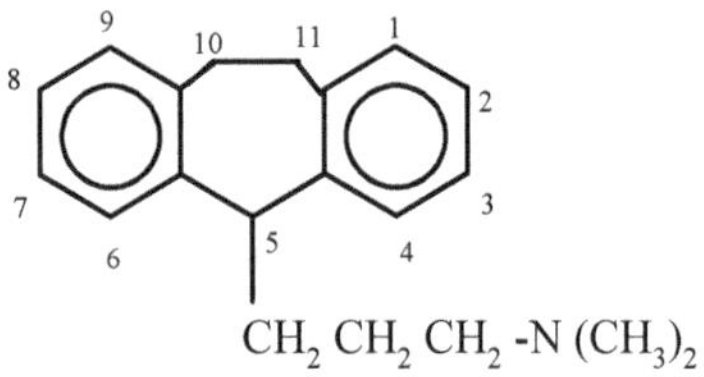

9
10
11
1
8
2
7
3
6
5
4
CH2 CH2 CH2 -N (CH3)2

5-[3-(Dimothyl amino) propyl] – dihydro – 5H – dbenz [b,1] ozopine

Synthesis

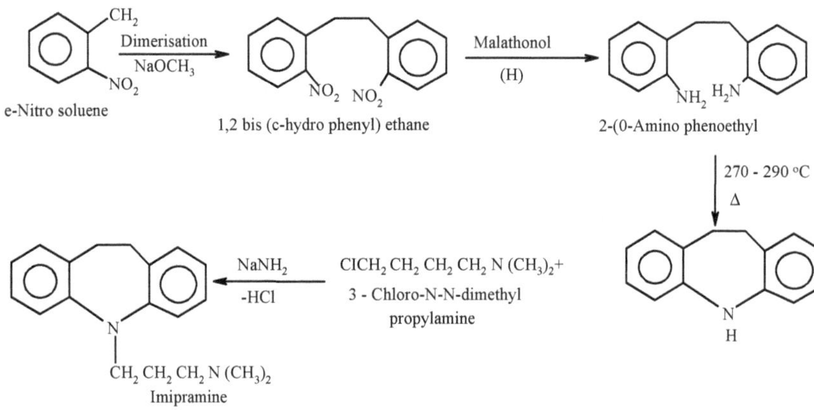

CH2
Dimerisation
NaOCH3
NO2
e-Nitro soluene
NO2 NO2
1,2 bis (c-hydro phenyl) ethane
Malathonol
(H)
NH2 H2N
2-(0-Amino phenoethyl
270 - 290 °C
Δ
N
H
ClCH2 CH2 CH2 CH2 N (CH3)2+
3 - Chloro-N-N-dimethyl
propylamine
NaNH2
-HCl
N
CH2 CH2 CH2 N (CH3)2
Imipramine

Amitriptyline

Amitriptyline is a psychoactive drug and pharmaceutical of the tricyclic antidepressant (TCA) chemical class which is used primarily as an antidepressant and anxiolytic agent. It is the most widely prescribed TCA and perhaps also the most efficient against depressive symptoms.

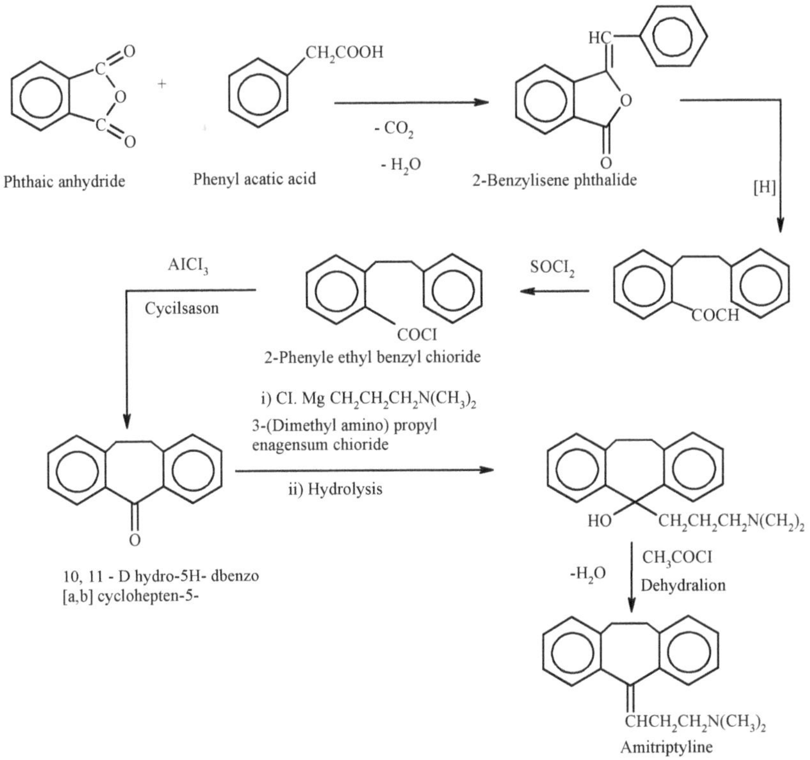

SAR for Tricyclic Antidepressants

1. The maximum antidepressant activity results on separation of the basic amino group from tricyclic nucleus by Propylene Bridge. The chain exceeding propyl group decrease activity.

2. 3-chloro derivative has less active than imipramine.

3. O-$(CH_3)_2$ derivative has equal potency.

4. Nuclear di substitution decreases the activity.

5. Piperazine propyl derivatives are found to be ineffective

For Dibenzo Cyclopentane derivatives

1. 3-Cl substitution enhance potency while a 3-CH_3 group diminish CNS depressant. Double bond between 10 &11 positions increases activity

2. The higher central ring homologue (octane) is more effective.

3. At position 11 the carbon is substituted by O, S, SO are clinically effective anti depressant. The bridged central ring also possesses powerful anti depressant activity.

Psychedelics

Psychedelics are agents which producing an increased awareness and enhanced perception of sensory stimuli. These are mind expanding drugs. These drugs can produce anxiety, fear, panic, hallucinations resembling to a psychosis. Hence they are called as hallucinogens and psychotomimetics.

Mechanism of action of Psychedelics

1. They induce or accelerate the production of hallucinogenic metabolites for noradrenaline.

2. They may cause charges in cerebral blood flow and permeability of cerebral capillaries.

3. They alters the levels of adrenal corticoids and thyroid hormones or changes in synthesis or metabolism of serotonin, nor epinephrine, acetylcholine or other potential transmitter.

4. Since serotonin is an inhibitors neurotransmitter, the removal of its inhibition could lead to behavioral changes.

5. These drugs may disrupt cerebral energy production or utilization in such a fashion that it alters the behaviors.

Anti-Depressants

An **antidepressant** is a psychiatric medication used to alleviate mood disorders, such as major depression and dysthymia. Drugs including the monoamine oxidase inhibitors (MAOIs), tricyclic antidepressants (TCAs), tetracyclic antidepressants (TeCAs), selective serotonin reuptake inhibitors (SSRIs), and serotonin-norepinephrine reuptake inhibitors (SNRIs) are most commonly associated with the term. These medications are among those most commonly prescribed by psychiatrists and other physicians, and their effectiveness and adverse effects are the subject of many studies and competing claims. Many drugs produce an antidepressant effect, but restrictions on their use have caused controversy and off-label prescription a risk, despite claims of superior efficacy.

Most typical antidepressants have a delayed onset of action (2–6 weeks) and are usually administered for anywhere from months to years. Despite the name, antidepressants are often used to treat other conditions, such as anxiety disorders, obsessive compulsive disorder, eating disorders, chronic pain, and some hormone-mediated disorders such as dysmenorrhea. Alone or together with anticonvulsants (e.g., Tegretol or Depakote), these medications can be used to treat attention-deficit hyperactivity disorder (ADHD) and substance abuse by addressing underlying depression. Also, antidepressants have been used for hypercytorism, with mixed reviews, and are sometimes used to treat snoring and migraines.

Benzo diazepines

Benzodiazepines and their derivatives are mainly used for sedative, hypnotic, tranquillizer, muscle relaxant and anti convulsant.

1. Diazepam

Diazepam

Synthesis of Diazepam

From this intermediate Diazepam can be obtained by various ways.

Method: I a

2. Chlor Diazepoxide

Chlordiazepoxide Hydrochloride

Synthesis of Chlor Diazepoxide

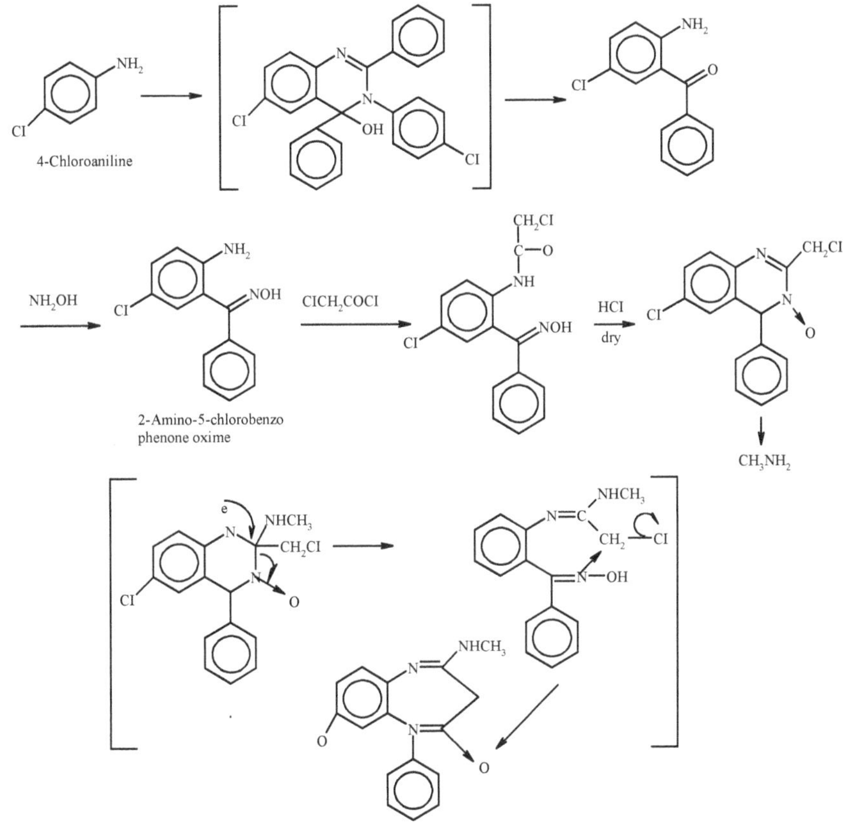
4-Chloroaniline
NH2OH
2-Amino-5-chlorobenzo phenone oxime
ClCH2COCl
HCl
dry
CH3NH2

Chlordiazepoxide (7.1)

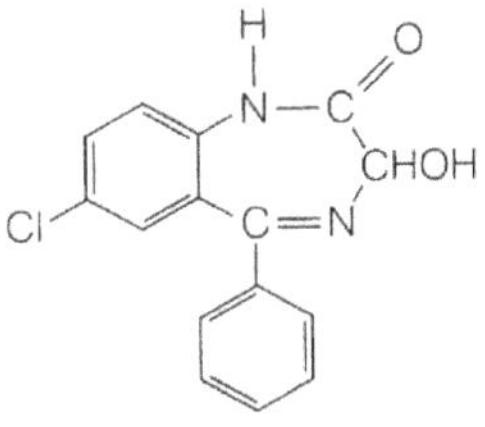

oxazepam

Meprobamate (marketed under the brand names **Miltown** by Wallace Laboratories, **Equanil** by Wyeth, and **Meprospan**) is a carbamate derivative which is used as an anxiolytic drug. It was the best-selling minor tranquilizer for a time, but has largely been replaced by the benzodiazepines.

Structure-Activity Relationship of Benzo Diazepines

1. All Benzodiazepines are used as CNS depressants are usually substituted with   a 5-aryl or 5- cyclo hexenyl groups.

2. The substituents present in position 1 to 3 do not affect the activity.

3. The N(4) usually substituted with a low electron density group.

4. The electron with drawing group present in the position 7 is required for hypnotic activity. Position 6, 8 & 9 should not be substituted.

5. A phenyl group at 5th position promotes the activity. An electron withdrawing group in ortho or di ortho position increases the activity. (Eg. Lorazepam, Flurazepam)

6. If 4, 5 double bond is saturated or shift to 3,4th position decreases the activity.

7. Alkyl substitution at 3rd position decreases the activity.

Mechanism of action

1. The mid brain reticular activating system which is responsible for the maintenance of wakefulness is depressed by benzodiazepines.

2. The anti anxiety activity of benzodiazepines may be attributed to the depressant action of these drugs on the mechanism which evoke anxiety and aggression.

3. Like GABA, the benzodiazepines cause either pre synoptic or post synoptic inhibition in poly synaptic neuronal pathways in CNS, affecting the various turnover of various neurotransmitters in the brain, therefore interfere the transmission process

Antispasmodics

An **antispasmodic** (synonym: **spasmolytic**) is a drug or an herb that suppresses spasms. These are usually caused by smooth muscle contraction, especially in tubular organs. The effect is to prevent spasms of the stomach, intestine or urinary bladder.

Cholinergic Blocking Agents Cholinergic Receptor Antagonists (or)
Anti Spasmodics

The anti cholinergic drugs are agents that inhibit the effect on Acetyl choline (Ach) released from post ganglionic parasympathetic nerve endings. They block muscarinic action of Ach including smooth muscle contraction and exocrine gland secretion. Because of the ability to relax smooth muscles, they are also referred as antispasmodics. The cholinergic blocking agents that have high affinity to the receptors may decrease the no. of available free receptors and the efficiency of the endogenous neuro transmitter like Ach. Anti muscarinic drugs act by competitive antagonism of Ach binding to muscarinic receptors.

Classification

1. ***Solanaceous alkaloids and Analogues*** : Atropine Sulfate, Hyoscyamine sulfate, Scopalamine HBr, Homatropine HBr, Ipratropium bromide.
2. ***Amino alcohol esters*** : Cyclopentolate. HCl, Clidinium bromide, Dicyclomine HCl, Glycopyrrolate, Methanthelin bromide, Propanthelin bromide, Mepenzolate.
3. ***Amino Alcohols*** : Biperidine HCl, Procyclidine HCl
4. ***Amino alcohol ethers*** : Benztropine mesylate, Orphenadrine
5. ***Amino amide*** : Tropicamide, Isopropamide iodide
6. ***Miscellaneous*** - ethopropazine.

For more details refer Anti Cholinergics in Unit 1

Dicyclomine is an anticholinergic that blocks muscarinic receptors. Dicyclomine is known as dicycloverine hydrochloride in the UK. Dicycloverine has 72 per cent of the anti-muscarinic power of atropine. It was invented in the United States in 1947.

Dicyclomine Hydrochloride

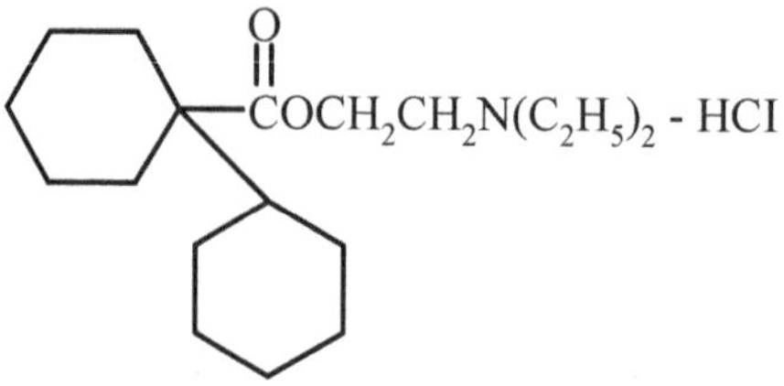
COCH2CH2N(C2H5)2 - HCI

2 - (Diethyl amino) cthyl bicyclohexyl - 1 carboxylate hydrochloride

Syntheshis

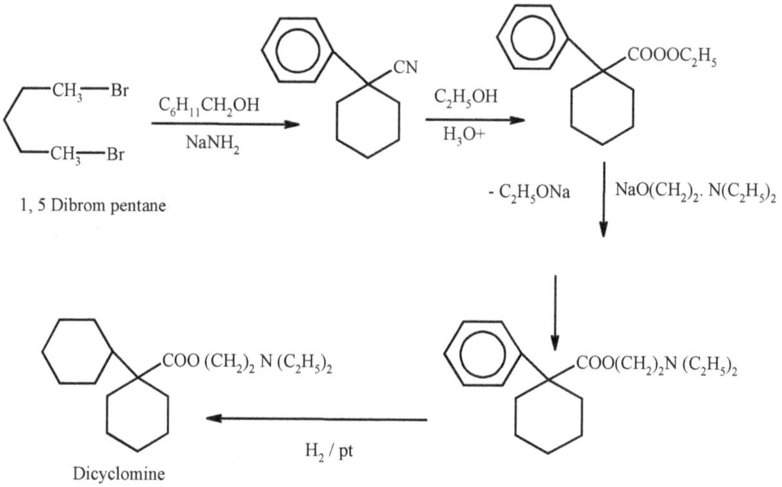
CH3—Br
CH3—Br
1, 5 Dibrom pentane
C6H11CH2OH
NaNH2
CN
C2H5OH
H3O+
COOOC2H5
- C2H5ONa
NaO(CH2)2. N(C2H5)2
COO(CH2)2N (C2H5)2
H2 / pt
COO (CH2)2 N (C2H5)2
Dicyclomine

Methantheline Bromide

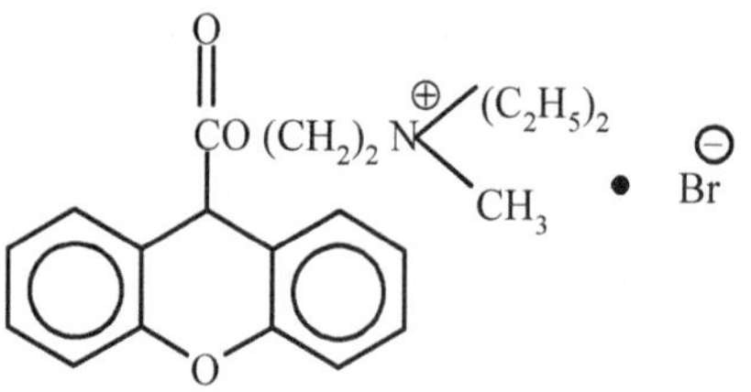
CO (CH2)2 N
(C2H5)2
CH3
Br
O

Anti ulcer Drugs

Antiulcer drugs are a class of drugs, exclusive of the antibacterial agents, used to treat ulcers in the stomach and the upper part of the small intestine.

Purpose

Recurrent gastric and duodenal ulcers are caused by *Helicobacter pylori* infections, and are treated with combination treatments that incorporate antibiotic therapy with gastric acid suppression. Additionally, bismuth compounds have been used. The primary classes of drugs used for gastric acid suppression are the proton pump inhibitors, omeprazole, lansoprazole, pantoprazole and rabeprazole. The H-2 receptor blocking agents, cimetidine, famotidine, nizatidine, and ranitidine have been used for this purpose, but are now more widely used for maintenance therapy after treatment with the proton pump inhibitors. Sucralfate, which acts by forming a protective coating over the ulcerate lesion, is also used in ulcer treatment and may be appropriate for patients in whom other classes of drugs are not indicated, or those whose gastric ulcers are caused by non-steroidal anti-inflammatory drugs (NSAIDs) rather than *H. pylori* infections.

Description

The proton pump inhibitors block the secretion of gastric acid by the gastric parietal cells. The extent of inhibition of acid secretion is dose related. In some cases, gastric acid secretion is completely blocked for over 24 hours on a single dose. In addition to their role in treatment of gastric ulcers, the proton pump inhibitors are used to treat syndromes of excessive acid secretion (Zollinger-Ellison Syndrome) and gastroesophageal reflux disease (GERD).

Histamine H-2 receptor blockers stop the action of histamine on the gastric parietal cells, inhibiting the secretion of gastric acid. These drugs are less effective than the proton pump inhibitors, but may achieve a 75-79% reduction in acid secretion. Higher rates of acid inhibition may be achieved when the drug is administered by the intravenous route. The H-2 receptor blockers may also be used to treat heartburn and hypersecretory syndromes. When given before surgery, the H-2 receptor blockers are useful in prevention of aspiration pneumonia.

Sucralfate (Carafate), a substituted sugar molecule with no nutritional value, does not inhibit gastric acid, but rather, reacts with existing stomach acid to form a thick coating that covers the surface of an ulcer, protecting the open area from further damage. A secondary effect is to act as an inhibitor of the digestive enzyme pepsin. Sucralfate does not bind to the normal stomach lining. The drug has been used for prevention of stress ulcers, the type seen in patients exposed to physical stress such as burns and surgery. It has no systemic effects

Omeprazole

5-Methoxy - 2 -(((4-methoxy - 3, 5-dimethyl - 2 pyridinyl) methyl) sulfinyl)-H-benzimidazole

Omeprazole is a proton pump inhibitor used in the treatment of dyspepsia, peptic ulcer disease (PUD), gastroesophageal reflux disease (GORD/GERD) and Zollinger-Ellison syndrome. It was first marketed in the US in 1989 by AstraZeneca under the brand names **Losec** and **Prilosec**, and is now also available from generic manufacturers under various brand names. AstraZeneca markets omeprazole as **Losec**, **Antra**, **Gastroloc**, **Mopral**, **Omepral**, and **Prilosec**. Omeprazole is marketed as **Zegerid** by Santarus, **Prilosec OTC** by Procter & Gamble and **Zegerid OTC** by Schering-Plough. Omeprazole is one of the most widely prescribed drugs internationally and is available over the counter in some countries. Prilosec contains the active ingredient omeprazole and Prilosec OTC contains the active ingredient omeprazole magnesium

Synthesis

Use in Helicobacter pylori eradication

Omeprazole is combined with the antibiotics clarithromycin and amoxicillin (or metronidazole in penicillin-hypersensitive patients) in the 7-14 day eradication triple therapy for *Helicobacter pylori*. Infection by *H. pylori* is the causative factor in the majority of peptic and duodenal ulcers.

Ranitidine

N-[2-[[[5-(Dimethylamino) methyl]-2- furany] thio] ethyl] -n'-methyl - 2- nitro -1, 1 - ethenediamine

Ranitidine hydrochloride is a histamine H_2-receptor antagonist that inhibits stomach acid production. It is commonly used in treatment of peptic ulcer disease (PUD) and gastroesophageal reflux disease (GERD). Ranitidine is also used alongside fexofenadine and other antihistamines for the treatment of skin conditions such as hives. Ranitidine HCl is marketed under the brand name **Zinetac** or **Zantac** (not to be confused with Xanax).

Synthesis

UNIT-IV

CENTRAL NERVOUS SYSTEM-III

- Antitussives
- Anticonvulsants
- CNS Stimulants
- Antiparkinsonian Drugs

CENTRAL NERVOUS SYSTEM-III

Anti Tussives

Anti tussives are the agents that are employed in the symptomatic control of cough by depressing cough centre situated in the medulla.

Coughing is a protective mechanism through which foreign materials, irritants and secretions are cleared from the respiratory tract.

Anti tussives can act either by raising threshold of the cough centre or by reducing the no of impulses transmitted to the cough centre from the peripheral receptors.

Classification

1. **Centrally Acting Anti Tussives**

   It affects the cough centre in the medulla.

   Example- Dextromethorphan Hydrobromide, Pholcodine, Noscapine, Carbeta   pentane

2. **Peripherally Acting Anti Tussives**

   It acts at the receptor level in the respiratory tract.

   Example- Benzonatate

Dextromethorphan is an antitussive drug. It is one of the active ingredients used to prevent coughs in many over-the-counter cold and cough medicines. Dextromethorphan has also found other uses in medicine, ranging from pain relief to psychological applications. It is sold in syrup, tablet, spray, and lozenge forms manufactured under several different brand names and generic labels. In its pure form, dextromethorphan occurs as a white powder. Dextromethorphan may be modestly effective in decreasing

Dextromethorphan Hydrobromide

3 - Methoxy -17- methyl - 9α, 13α, 14 α -morphinan hydrobromide

cough associated with an upper respiratory tract infection in adults. In children however it has not been found to be effective. When exceeding label-specified maximum dosages, dextromethorphan acts as a dissociative psychedelic drug. Its mechanism of action is as an NMDA receptor antagonist, producing effects similar to those of the controlled substances ketamine and phencyclidine (PCP).

Mechanism of Action

Dextromethorphan control the cough by depressing the cough centre in the medulla. The potency is almost one-half of the codeine. But this drug does not producing addiction even after the usage of large doses for prolonged duration like codeine. Benzonatate reducing the cough reflux at its source by anesthetizing the stretch receptors located in the respiratory passages and lungs.

Dextromothorphan

Dextrobane

3-Methanomorphinan

Glucoronide and
Sulfate conjugates

Glucoronide and
Sulfate conjugates

5-Hydroxy morphlan

Cramiphen

Diethylaminoethyl-1-phenylcyclopentane-1-carboxylate   hydrochloride;
an antitussive.

$CH_3 (CH_2)_3 NH$ — — $CO (CH_2.CH_2.O)_9 OCH_3$

4-(Butylamino) benzoic acid - 3, 6, 9, 12, 15, 18, 21, 24, 27- nonaoxaoctacos-1-yl ester.

Anti Convulsant Drugs (or) Anti Epileptic Drugs

Epilepsy is a collective term for a group of chronic CNS disorders having
in common, sudden and transitory seizers of loss or disturbance of
consciousness with characteristic body movements (convulsions) and
some times with autonomic hyperactivity. The principal types of epilepsy
are:

1. *Grandmal epilepsy*: It is normally characterized by complete loss of consciousness followed by transient muscular rigidity and clonic convulsions in all voluntary muscles.

2. *Petitmal epilepsy*: It is usually momentary loss of consciousness. There is free of convulsions and occasionally blinking movements of eyelids and jerking movements of the head and arms.

3. *Psychomotor epilepsy*: It is characterized by attacks without convulsions lasting from 2-3 minutes. It displays mental apathy and sudden irrational and destructive attitude.

4. *Myoclonic seizers*: It is characterized by jerky muscular movements of head, limbs or body as such. The duration of attack remains near about one second and reappears at about 5 secs intervals for 1 min. It is rapid rhythmic movement.

Anti Convulsants: Anti Convulsant drugs are also termed as antiepileptic drugs, are drugs which selectively depress the CNS and prevent or control the epileptic seizers. The drugs are adequate and impressive control and management of CNS disorders essentially characterized by recurrent transient attacks of disturbed brain function which ultimately give rise to motor, (convulsive), sensory, (seizures) and psychic sequence of events.

Classification

1. *Barbiturates*: Phenobarbital, Mephobarbital, Metharbital

2. *Hydantoins*: Phenytoin, Mephenytoin, Ethotoin

3. *Oxazolidine dione*: Trimethadione, Paramethadione

4. *Succinimides*: Phen suximide, Methsuximide, Ethosuximide

5. *Urea derivatives*: Phenacemide, Carbamezepine

6. *Benzodiazepines*: Clonazepam, Diazepam, Chlorazepate

7. *Miscellaneous*: Primidone, Valproic acid, Gabapectin, Felbamate.

1. *Barbiturates*

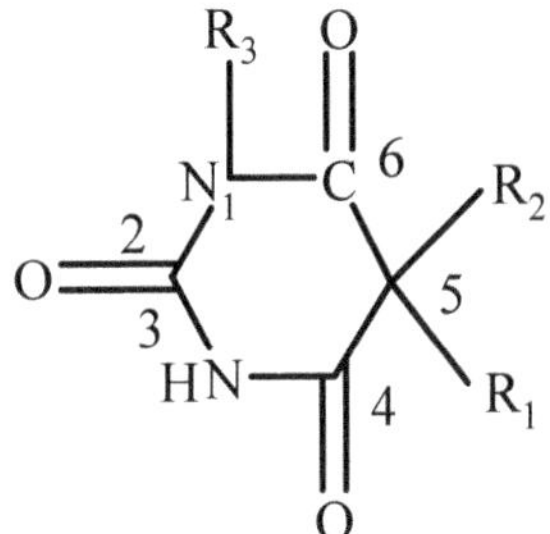

General Structure

Name	R_1	R_2	R_3
Phenobarbitone	C_2H_5	C_2H_5	H
Mephobarbitone	C_2H_5	C_2H_5	CH_3
Metharbital	CH_3	CH_3	CH_3

2. *Hydantoins*

2, 4 - Imidazolidinedione

S. No	Generic Name	Substituents		
		R_2	R_3	R_1
1.	Phenytoin	C_6H_5	C_6H_5	H
2.	Mephenytoin	C_6h_5	C_2H_5	CH_3
3.	Ethotoin	C_6H_5	H	C_2H_5

Synthesis of Phenytoin

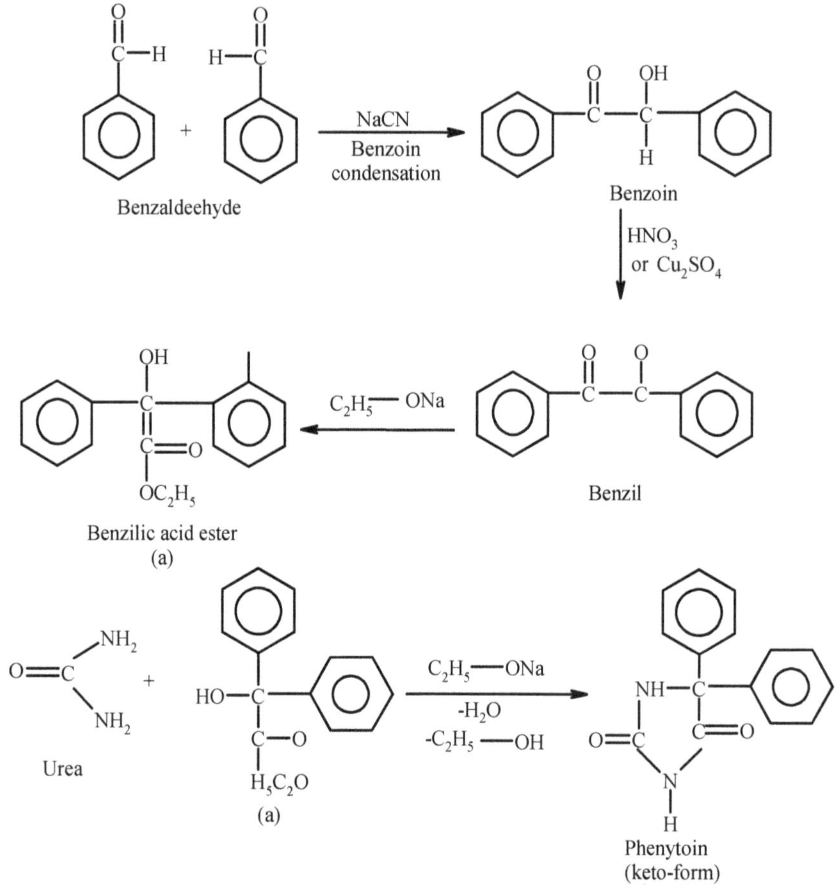

Benzaldeehyde

NaCN
Benzoin condensation

Benzoin

HNO_3
or Cu_2SO_4

C_2H_5—ONa

Benzil

Benzilic acid ester
(a)

OC_2H_5

Urea

C_2H_5—ONa

$-H_2O$

$-C_2H_5$—OH

(a)

Phenytoin
(keto-form)

3. *Oxazolidinediones*

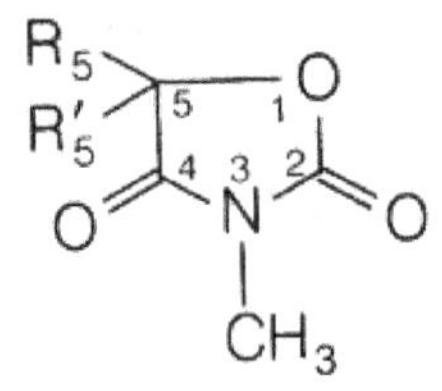

Trimethadione $R_5 = R'_5 = CH_3$
Paramethadione $R_5 = CH_3$; $R'_5 = C_2H_5$

4. *Succinimides*

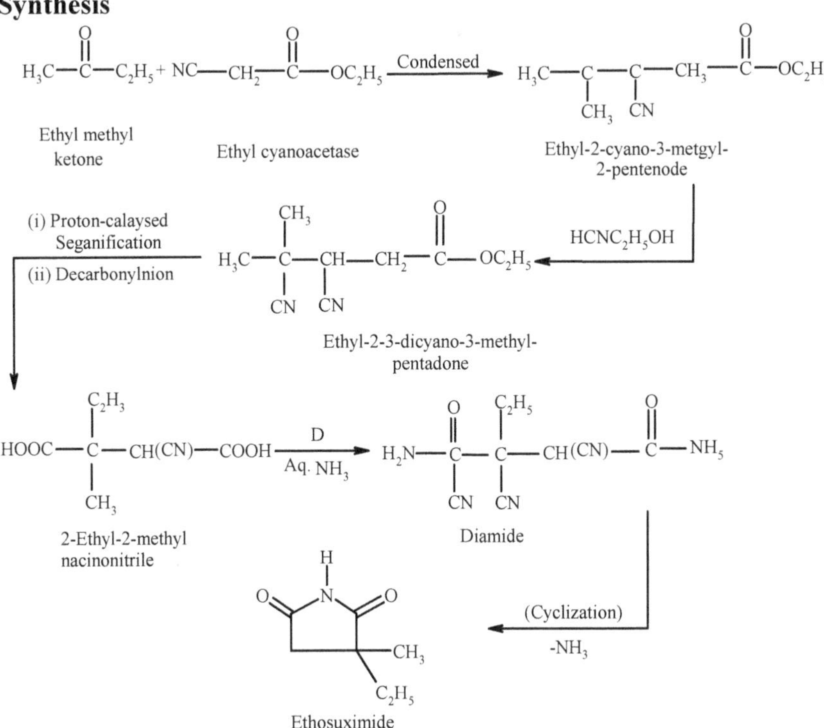

S.No	Generic Name	R	R'	R''
1.	Phensuximide	C_6H_5	H	CH_3
2.	Methsuximide	C_6H_5	CH_3	CH_3
3.	Ethosuximide	C_2H_5	CH_3	H

Synthesis

5. *Urea Derivatives*

   1. Phenacemide

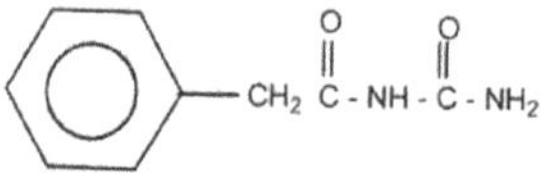

Phenyl acetyl urea

   2. Carbamazepine

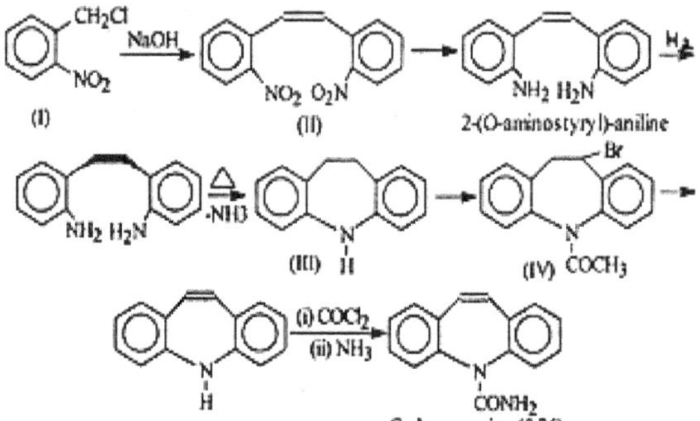

(I)   (II)   2-(O-aminostyryl)-aniline

(III)   (IV) COCH$_3$

Carbamazepine (8.26)

6. *Miscellaneous*

Valproic acid

$$CH_3CH_2CH_2CH\text{-}COOH$$

$$CH_3CH_2CH_2$$

2-n-Propyl pentanoic acid

Synthesis

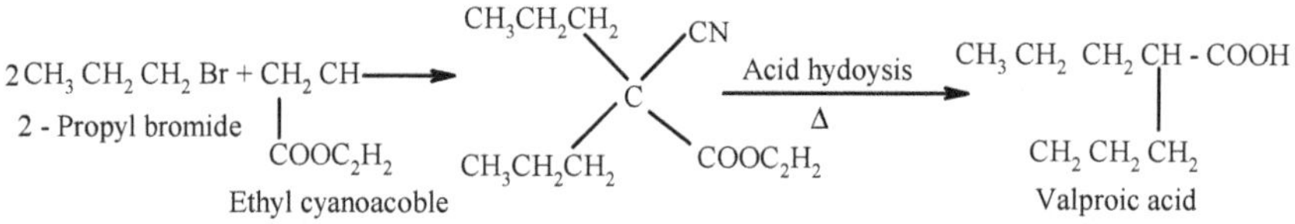

$$2\,CH_3\,CH_2\,CH_2\,Br + CH_2\,CH \longrightarrow$$

2 - Propyl bromide

COOC$_2$H$_2$

Ethyl cyanoacoble

Acid hydoysis, Δ

$$CH_3\,CH_2\,CH_2\,CH\text{-}COOH$$

$$CH_2\,CH_2\,CH_2$$

Valproic acid

Structure Activity Relationship

For Barbiturates

   1. Phenyl or other aromatic substituents at 5th position is essential for good activity

   2. 5, 5 diphenyl derivatives have less active than phanobarbitone.

   3. N-substitutions also increase the antiepileptic activity.

   4. 5, 5-dibenzyl barbiturates cause convulsions.

For Hydantoins

1. Phenyl or other aromatic substituents at 5th position is essential for good activity.

2. Alkyl substitution at 5th position may contribute to sedation.

3. Some Thio hydantoins, Dithio hydantoins and 1, 3-disubstituted hydantoins exhibit activity against chemically induced convulsions and ineffective against electroshock induced convulsions.

For Oxazolidine Diones

1. The nature of substituents at 5th position is essential for good activity.

2. The lower substituents tend towards anti-petitmal epileptic activity and aryl towards anti-grandmal epileptic activity.

3. N-alkyl substituents do not affect the activity, because the N-dealkylated metabolites are active anticonvulsant agents.

4. Alkylation of imido nitrogen is more active, because it increases partition coefficient and prevents the dissociation of imido nitrogen and more distribution to CNS.

For Succinimides

1. Meth suximide and phen suximide have phenyl substituents at 3rd position are active, ineffective against electroshock induced convulsions.

2. N-methylation decreases the activity against electroshock induced convulsions and more activity against chemically induced convulsions.

3. Alpha-methyl alkoxy phenyl succinimides and alkoxy benzyl succinimides were active anti convulsants.

For Benzodiazepines

1. The electron withdrawing groups at 7th position increases antiepileptic activity and electron donating groups at 7th position increases antiepileptic activity.

2. A phenyl group at 5th position is necessary for good activity. But the halogen substituents in phenyl group in ortho position increase the activity.

3. The electron withdrawing groups at ortho or di-ortho positions at 5-Phenyl group increases antiepileptic activity while any

substituent on meta or para position of 5-Phenyl group decreases antiepileptic activity.

4. Methyl substitution at 1st position increases the activity.

Mechanism of Action

1. The generation of seizures is due to excessive discharge of neurotransmitter in CNS.The anti convulsant drugs increase the level of serotonin in brain which causes non specific depression of CNS functions and controls the release of neurotransmitters.

2. Gama Amino Butyric Acid (GABA) levels in brain are also important to prevent the speed of seizures. So anti convulsant drugs increase the level of GABA in brain.

3. Anti convulsant activity of barbiturates is attributed to their ability to exert conformational rearrangement of oxidative enzymes essential for brain respiration.

4. Many carbonic an hydrase inhibitors have Anti convulsant activity by decreasing the cerebral respiration due to excess CO_2 depress the nerve function.

5. Anti convulsant drugs usually display various combined activities on the neuronal function such as act on ion channels and maintain the neuronal membrane, the resting potential having range of -50 mv to -80 mv between inside (K+) and outside (Na+ & Cl$^-$) of the cell.

6. GABA binds to GABA alpha and GABA beta receptors. The oscillation rhythms in epilepsy caused by GABA alpha receptors. Therefore the drugs potenciate GABA mediated inhibitors or to affect the GABA concentration in brain.

CNS Stimulants

The drugs that produce stimulation of central nervous system and enhancement in excitability of different portions of the brains or the spinal cord.

The CNS stimulants include analeptics, antidepressants, central sympathomimetic agents (Psychomotor stimulants). Some times CNS stimulants lead to convulsions so they are limited therapeutic value because of their convulsant activities and side effects.

Classification

1. *Analeptics*: Picrotoxin, Nikethamide, Etamivan, Pemoline Pentylene tetrazole

   (Pentetrazole), Doxopram, Bemegride, Strychnine.

2. *Methyl xanthines*: Caffeine, Theophylline, Theobromine, Aminophylline, Etophylline, Proxyphylline.

3. *Central sympathomimetic agents (Psychomotor stimulants)*: Amphetamine, Methamphetamine, Phentermine, Benz phentamine, Chlorpentermine, Ferfluramine, Chlortermine, Phenmetrazine, Phendimetrazine, Mazindol, Methyl phenidate.

4. *Mono amino oxidase Inhibitors (MAO – inhibitors*: Phenelzine, Isocarboxazid, Tranyl cypromine, pargyline, clorgyline.

5. *Tricyclic Antidepressants*: Imipramine, Desipramine, Amitryptyline, Nortriptyline, Protriptyline, Trimipramine, Doxepin, Maprotiline.

6. *Psychedelics*:

   1. *Indole ethyl amines* – Bufotenine, Psilocybin, Psilocyn

   2. *2 – Phenyl ethylamines* – Mescaline

   3. *Agents have both indolethylamine and phenyl ethylamine* – (+) Lysergic

      acid, Diethylamide (LSD).

   4. *Dissociative agents* – Phencyclidine (PCP)

   5. *Depressant-Intoxicant* – Tetra hydro cannabinol (THC)

I. **Analeptics:** Analeptics are agents which stimulate various areas of the central nervous system. These are mainly used for the treatment of respiratory depression resulting from overdose of depressant drugs. So these are used as respiratory stimulants. An excessive dose of analeptics may result a wide–spread stimulation of the brain that may ultimately cause convulsions.

    Mechanism of Action for Analeptics

    1. Some drugs block post synaptic inhibition (Strychmine) or pre synaptic inhibition (Picrotoxin).

    2. Some drugs acts as GABA ontagonist (Picrotoxins bemegride) or release prostoglandin and also decrease energy levels (rentetrazole)

Nikethamide

N, N – Diethyl nicotinamide

Nikethamide Synthesis

SAR for Central Sympathomimetic Agents

Any decrease in distance between aromatic ring and heterocyclic nitrogen decrease the activity.

- The branched CH_3 group or similar substitution is important for activity, since it provides resistance to enzymatic in activation by steric protection of the amino group.

- In phenidate series, activity is maximal at the methyl ester.

- In morpholine series aromatic substitutions and replacement of ring by heterocyclic groups decrease the activities.

Mechanism of Action

1. They inhibit reuptake mechanisms for several biogenic amines.
2. They enhance neuronal release of catecholamines.
3. They stimulate a-adrenergic receptor and inhibit mono amino oxidase is higher concentration.

II. Methyl Xanthines

Caffeine is a bitter, white crystalline xanthine alkaloid that is a psychoactive stimulant drug. Caffeine was discovered by a German chemist, Friedrich Ferdinand Runge, in 1819. He coined the term *kaffein*, a chemical compound in coffee, which in English became *caffeine.*

Santhine derivatives

Compound	R_1	R_3	R_2
Caffeine	CH_3	CH_3	CH_3
Theophylline	CH_3	CH_3	H
Theobromine	H	CH_3	CH_3
Etofyline	CH_3	CH;	$CH_3 CH_2 OH$
Proxyphyline	CH_3	CH_3	$CH_2 CH (OH) - CH_3$
Pentoxyphyline	$(CH_2)_4 CO CH_3$	CH_3	CH_3

Mechanism of Action

Caffeine readily crosses the blood–brain barrier that separates the bloodstream from the interior of the brain. Once in the brain, the principal mode of action is as a nonselective antagonist of adenosine receptors. The caffeine molecule is structurally similar to adenosine, and binds to adenosine receptors on the surface of cells without activating them (an "antagonist" mechanism of action). Therefore, caffeine acts as a competitive inhibitor.

Synthesis

Antiparkinsonian Agents

Parkinson's disease (also known as **Parkinson disease** or **PD**) is a degenerative disorder of the central nervous system that often impairs the sufferer's motor skills, speech, and other functions.

Parkinson's disease belongs to a group of conditions called movement disorders. It is characterized by muscle rigidity, tremor, a slowing of physical movement (bradykinesia) and, in extreme cases, a loss of physical movement (akinesia). The primary symptoms are the results of decreased stimulation of the motor cortex by the basal ganglia, normally caused by the insufficient formation and action of dopamine, which is produced in the dopaminergic neurons of the brain. Secondary symptoms may include high level cognitive dysfunction and subtle language problems. PD is both chronic and progressive.

Anti-Parkinsonian Drugs Include

Anticholinergic Agents; COMT Inhibitors; Dopamine Agonists; Levodopa; MAO-B Inhibitors.

Treatment

Parkinson's disease is a chronic disorder that requires broad-based management including patient and family education, support group services, general wellness maintenance, physiotherapy, exercise, and nutrition. One could consult an occupational therapist on a broad range of methods, therapy, and assistive equipment to make Parkinson's more manageable in the areas of personal care, productivity and leisure. At present, there is no cure for PD, but medications or surgery can provide relief from the symptoms.

Levodopa

The most widely used form of treatment is L-dopa in various forms. L-dopa is transformed into dopamine in the dopaminergic neurons by L-aromatic amino acid decarboxylase (often known by its former name dopa-decarboxylase). However, only 1-5% of L-DOPA enters the dopaminergic neurons. The remaining L-DOPA is often metabolised to dopamine elsewhere, causing a wide variety of side effects. Due to feedback inhibition, L-dopa results in a reduction in the endogenous formation of L-dopa, and so eventually becomes counterproductive.

Carbidopa and benserazide are dopa decarboxylase inhibitors. They help to prevent the metabolism of L-dopa before it reaches the dopaminergic neurons and are generally given as combination preparations of carbidopa/levodopa (co-careldopa) (e.g. Sinemet, Parcopa) and benserazide/levodopa (co-beneldopa) (e.g. Madopar). There are also controlled release versions of Sinemet and Madopar that spread out the effect of the L-dopa. Duodopa is a combination of levodopa and carbidopa, dispersed as a viscous gel. Using a patient-operated portable pump, the drug is continuously delivered via a tube directly into the upper small intestine, where it is rapidly absorbed. Another drug, Stalevo (carbidopa, levodopa and entacapone), is also available for treatment.

Tolcapone inhibits the COMT enzyme, thereby prolonging the effects of L-dopa, and so has been used to complement L-dopa. However, due to its possible side effects such as liver failure, it's limited in its availability. A similar drug, entacapone has not been shown to cause significant alterations of liver function and maintains adequate inhibition of COMT over time.

Dopamine Agonists

The dopamine agonists bromocriptine, pergolide, pramipexole, ropinirole, piribedil, cabergoline, apomorphine, and lisuride are moderately effective. These have their own side effects including those listed above in addition to somnolence, hallucinations and/or insomnia. Several forms of dopamine agonism have been linked with a markedly increased risk of problem gambling. Dopamine agonists initially act by stimulating some of the dopamine receptors. However, they cause the dopamine receptors to become progressively less sensitive, thereby eventually increasing the symptoms.

Dopamine agonists can be useful for patients experiencing on-off fluctuations and dyskinesias as a result of high doses of L-dopa. Apomorphine can be administered via subcutaneous injection using a small pump which is carried by the patient. A low dose is automatically administered throughout the day, reducing the fluctuations of motor symptoms by providing a steady dose of dopaminergic stimulation. After an initial "apomorphine challenge" in hospital to test its effectiveness and brief patient and primary caregiver (often a spouse or partner), the latter of whom takes over maintenance of the pump. The injection site must be changed daily and rotated around the body to avoid the formation of nodules. Apomorphine is also available in a more acute dose as an autoinjector pen for emergency doses such as after a fall or first thing in the morning. Nausea and vomiting are common, and may require domperidone (an antiemetic).

MAO-B Inhibitors

Selegiline and rasagiline reduce the symptoms by inhibiting monoamine oxidase-B (MAO-B). MAO-B breaks down dopamine secreted by the dopaminergic neurons, so inhibitting it will result in inhibition of the breakdown of dopamine. Metabolites of selegiline include L-amphetamine and L-methamphetamine (not to be confused with the more notorious and potent dextrorotary isomers). This might result in side effects such as insomnia. Use of L-dopa in conjunction with selegiline has increased mortality rates that have not been effectively explained. Another side effect of the combination can be stomatitis. One report raised concern about increased mortality when MAO-B inhibitors were combined with L-dopa; however subsequent studies have not confirmed this finding. Unlike other non selective monoamine oxidase inhibitors, tyramine-containing foods do not cause a hypertensive crisis.

Carbidopa

Carbidopa (Lodosyn) is a drug given to people with Parkinson's disease in order to inhibit peripheral metabolism of levodopa.

Pharmacology

Carbidopa inhibits aromatic-L-amino-acid decarboxylase (DOPA Decarboxylase or DDC), an enzyme important in the biosynthesis of L-tryptophan to serotonin and in the biosynthesis of L-DOPA to Dopamine (DA).

Along with carbidopa, other DDC inhibitors are benserazide (Ro-4-4602), difluromethyldopa, and α-methyldopa.

Uses

Used in tandem with L-DOPA (trade name levodopa, a dopamine precursor converted in the body to dopamine), it increases the plasma half-life of levodopa from 50 minutes to 1 1/2 hours. CarbiDOPA cannot cross the blood brain barrier, so it inhibits only peripheral DDC. It thus prevents the conversion of L-DOPA to dopamine peripherally. This reduces the side effects caused by dopamine on the periphery, as well as increasing the concentration of L-DOPA and dopamine in the brain.

The combination of carbidopa/levodopa) carries the brand names of **Sinemet, Parcopa** and **Atamet**; whilst **Stalevo** is a combination with entacapone, which enhances the bioavailability of carbidopa and levodopa.

Carbidopa is also used in combination with 5-HTP, a naturally occurring amino acid which is a precursor to the neurotransmitter serotonin and an intermediate in tryptophan metabolism. Carbidopa prevents 5-HTP's metabolism in the liver, which can lead to elevated levels of serotonin in the bloodstream. Research shows that co-administration of 5-HTP and carbidopa greatly increases plasma 5-HTP levels. Without the use of carbidopa, there is a significant risk of heart valve disease when taking 5-HTP, due to serotonin's affect on the heart. In Europe, 5-HTP is prescribed with **carbidopa** to prevent the conversion of 5-HTP into serotonin until it reaches the brain.

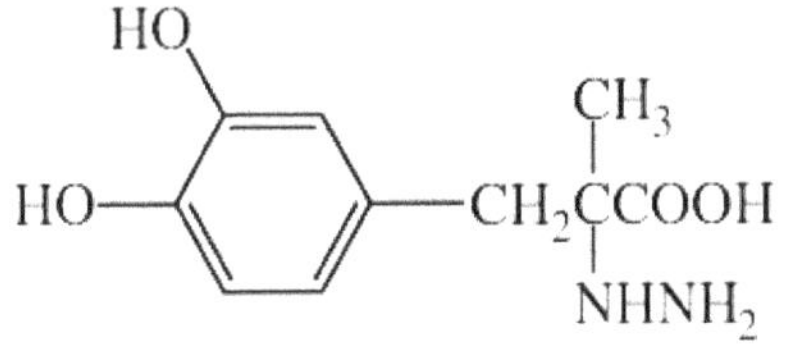

CARBIDOPA

2*S*)-3-(3,4-dihydroxyphenyl)-2-hydrazino-2-methylpropanoic acid

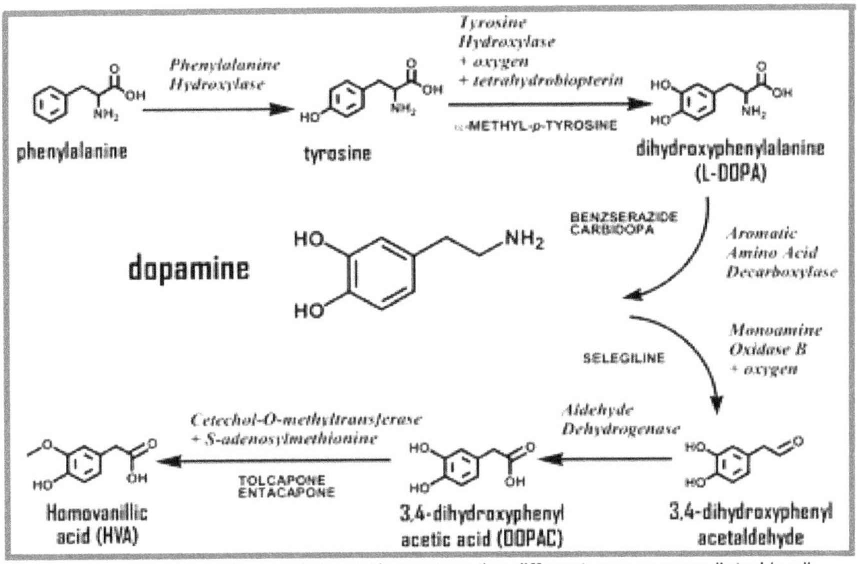

Enzymes are listed in italics, and drugs that act on the different enzymes are listed in all capital letters.

Fig. 4.1 Biochemical pathways of dopamine metabolism

L-DOPA or Levodopa

L-DOPA (**L-3,4-d**ihydroxyphenylalanine; **Levodopa**; **Sinemet**, **Parcopa**, **Atamet**, **Stalevo**, **Madopar**, **Prolopa**, etc) is a naturally-occurring dietary supplement and psychoactive drug found in certain kinds of food and herbs (e.g. *Mucuna pruriens*, or velvet bean), and is synthesized from the essential amino acids L-phenylalanine (PHE) and L-tyrosine (TYR) in the mammalian body and brain. L-DOPA is the precursor to the neurotransmitters dopamine, norepinephrine (noradrenaline), and epinephrine (adrenaline) collectively known as catecholamines. Aside from its natural and essential biological role, L-DOPA is also used in the clinical treatment of Parkinson's disease (PD) and dopamine-responsive dystonia.

(*S*)-2-amino-3-(3, 4-dihydroxyphenyl) propanoic acid

UNIT-V
AUTOCOIDS

- Antihistaminics (a) H1 antagonists (b) H2 antagonists
- 5-HT, 5HT-antagonist
- Angiotensin antagonist
- Prostaglandin

AUTOCOIDS

ANTI Histaminic Agents

Histamine is a beta imidazole ethylamine derivative which is present in essentially all mammalian tissues. In the living organism histamine is synthesized from the naturally occurring a-amino acid, histidine by the loss of a carboxyl group through bacterial or enzymatic decarboxylation.

$$N\text{---}C\text{-}CH_2.COOH \quad \xrightarrow[\text{-}CO_2]{\text{Decarboxylation}} \quad N\text{---}C\text{-}CH_2.CH_2.NH_2$$

Histitne → Histamine

Histamine is produced naturally by human system and released in response to tissue damage. In human beings, histamine cause immediate allergic and inflammatory response cause gastric acid release and function as a CNS neurotransmitter.

Systemically, histamine contracts smooth muscle of lungs and gastro intestinal system and cause vasodilatation, low B.P and increase heart rate. It causes symptoms such as itching, sneezing, watery eye and running nose.

Histamine receptors

The physiological effects of histamine are mediated by specific cell-surface receptors. These receptors are divided into three types.

1. H1- receptors

It is found in smooth muscles of intestine, bronchi, blood vessels, adrenal medulla, endothelial cell and lymphocytes. Histamine H1 receptors are G-protein linked receptors. It is sequence of 491 amino acids residue. The third and fifth membrane domain is responsible for binding histamine.H1 - receptors mediate smooth muscle contraction, increased vascular permeability, pruritus, prostaglandin, decreased artrioventricular conduction time accompanied by tachycardia and activation of vagal reflexes.

2. H2 - receptors

They are located on the cell membrane of acid secreting cells of gastric mucosa and mediate the gastric acid secretary actions of histamine. The physiological effects of H2 - receptor ligands are mediated by a stimulatory G- protein coupled receptor which activates adenylate cyclase / cyclic AMP intra cellular second messenger. The TM3 aspartate and aspartate and threonine residue in TM5 is responsible for binding histamine.

3. H3 - receptors

It is presynoptic receptor that influences the release of histamine and other neuro transmitters from neurons.

Anti Histamines

Anti histamines are drugs which inhibit the action of histamine by competitively blocking the histamine receptors.

A. Histamine H1 Receptor Antagonist

Classification

I. **First generation anti histamines**

1. **Amino alkyl ethers:** Diphen hydramine HCl, Dimenhydrinate, Bromodiphenhydramine HCl, Doxylamine succinate, Carbinoxamine maleate, Clemastine fumerate, Diphenyl pyraline HCl.

2. **Ethylene diamines:** Tripelennamine HCl, Pyrilamine maleate, Metha pyrilene HCl, Thonzylamine HCl, Antazoline PO4.

3. **Piperazine derivatives:** Cyclizine HCl, Chlorcyclizine HCl, Meclizine HCl, Buclizine HCl.

4. **Propylamine derivatives (Mono amino propyl derivatives):** Pheniramine maleate, Chlor Pheniramine maleate, Triprolidine HCl, Phenindamine tartarate, Pyrrobutamine PO4, Dimethindene maleate, Dexchlorpheniramine maleate, Brompheniramine maleate, Dex brompheniramine maleate.

5. **Phenothiazine derivatives:** Promethazine HCl, Trimeprazine tartarate, Methdilazine.

6. **Dibenzocyloheptene derivatives:** Cyproheptadine HCl, Azatadine maleate.

II. Second generation H1 antagonist

Terfenadine, Astemizole, Loratadine, Cetirizine, Acrivastine.

III. Inhibition of Histamine release (Mast cell stabilizers)

Cromolyn sodium, Nedocromil sodium.

B. Histamine H2 Receptor Antagonist

Cimetidine, Famotidine, Ranitidine,

C. Other Antiulcer Agents

Omeprazole, Lansoprazole, Pantoprazole, Rabeprazole.

1. Amino alkyl ethers

The General formula is

$$Ar_2\text{-}CH\text{-}O\text{-}CH_2CH_2 \ N \ (R)_2$$

Diphenhydramine

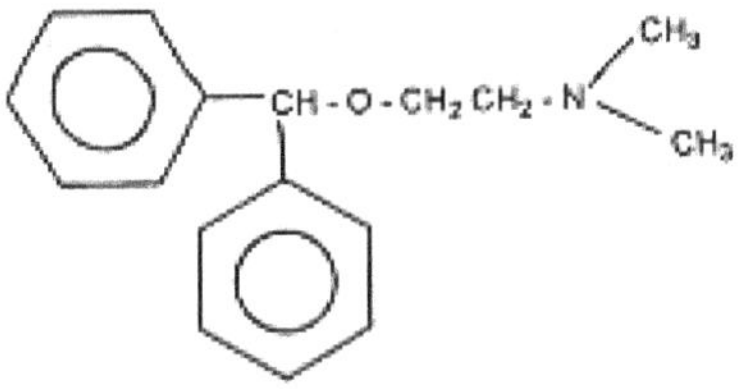

2- (Diphenylmethoxy)-N,N-dimethyl ethonamine

Synthesis

Bromodiphenhydramine

SAR for amino alkyl ether

The aromatic group may be phenyl or substituted phenyl or heterocyclic for good   antihistaminic activity.

- P–substituted aromatic groups have good activity but O-substitution in aryl groups loss the activity.

- Removal of alpha1–methyl group and insertion of chlorine in to Para position of the phenyl ring in Doxylamine enhanced activity.

- Substitution of 2–thionyl for 2–pyridyl group decreased the activities.

- Compounds with an asymmetric carbon atom, the Dextro isomer is more active.

- If double bond is introduced between a, b carbon atoms of the propyl chain drowsiness will be developed.

Ethylene Diamines

The General formula is

$$(Ar)_2\ N\text{-}CH_2CH_2\text{-}N\ (CH3)_2$$

Pyrilamine (Mepyramine)

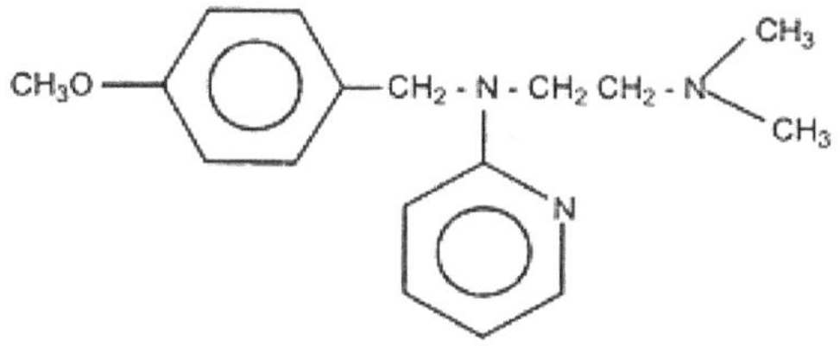

2- {[2-(Dimethyl amino) -ethyl] -(p-methoxy benzyl) - amino] pyridine}.

SAR for ethylene Diamines

$$(Ar)_2\ N\ \text{-}\ CH_2CH_2\ \text{-}N\ (CH3)_2$$

1. One of the aryl group is 2–pyridinyl system is more significant anti histaminic activity.

2. Substitution of P–methoxy (Pyrilamine), chloro (Chlor pyramine) or bromo (Brom tripelenamine) group enhances the activity.

3. The two tertiary nitrogens are separated by two carbon chain for good activity. Extension or branching of this chain decrease activity.

4. The tertiary nitrogen may be a part of heterocyclic ring (antazoline) also has good activity.

3. Piperazine Derivatives

Meclizine

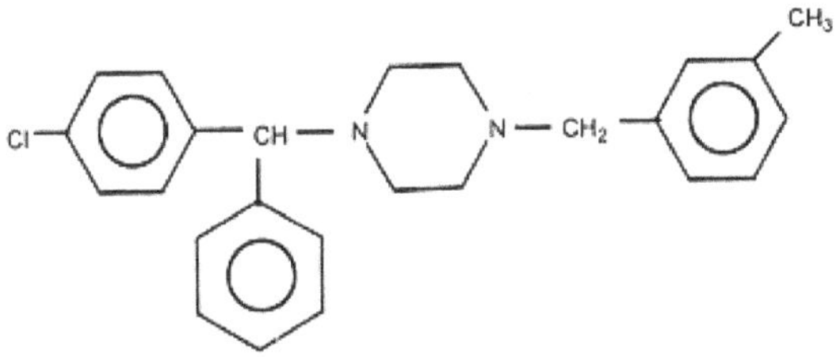

1-(p-Chloro -α - phenyl benzyl) - 4 - (m-methyl benzyl) piperazine.

SAR for Piperazine Derivatives

1. P-Substitution of any one aryl group by chlorine enhances the anti histaminic activity.

2. The both nitrogen atoms of Piperazine are aliphatic and have basicities for good activities.

3. The R- group may be methyl or aralkyl for good activity.

Propylamine Derivatives

The General formula is

$$(Ar)_2 \text{ CH } CH_2CH_2\text{-N } (R)_2$$

Pheniramine

2- [α - (2-Dimethyl amino ethyl)- benzyl]- pyridine.

SAR for Propylamines

1. One of the aryl group may be 2–Pyridinyl group is more significant antihistaminic activity.

2. Introduction of chlorine in P- position of benzyl group has 20 times more potent than unsubstituted compounds.

3. These drugs have an asymmetric carbon atom, the Dextro isomers exhibiting the greater potency.

4. In the unsaturated derivatives, the transisomers are more active.

5. The tertiary nitrogen may be a part of heterocyclic ring (Triprolidine) has greatest activity.

V. Phenothiazine derivatives

Promethazine

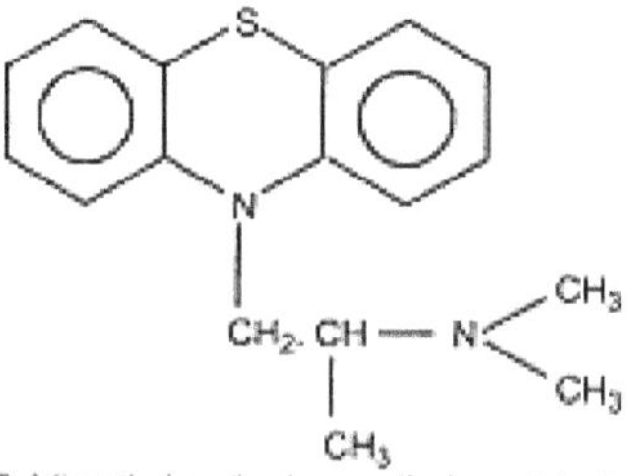

10- [2-(dimethyl amino) propyl] phenothiazine.

Synthesis

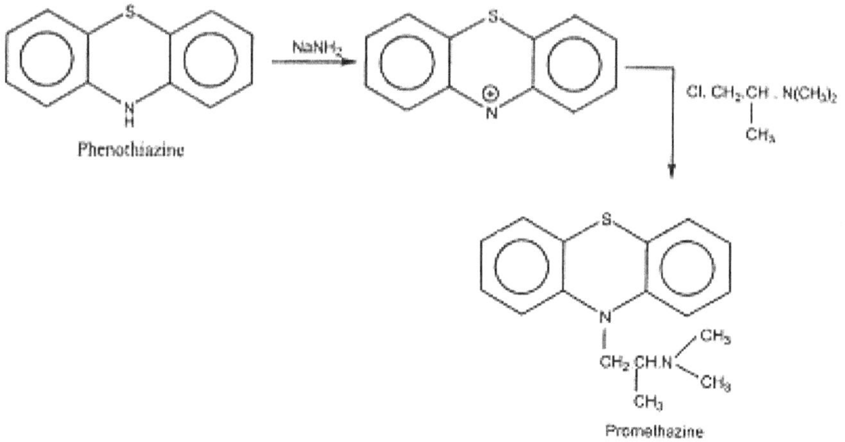

Phenothiazine

NaNH₂

Cl. CH₂.CH . N(CH₃)₂
CH₃

Promethazine

SAR for Phenothiazines

1. The side chain of phenothiazines contain two or three carbon atoms, branched alkyl chain between ring system and terminal nitrogen atom gives good anti histaminic activity.

2. The phenothiazines with a 3 carbon bridge between nitrogen atoms are more potent *in vitro.*

3. The 30 nitrogen of side chain may be a part of heterocyclic ring also shows good activity.

6. Di benzo cycloheptanes

Cyproheptadine

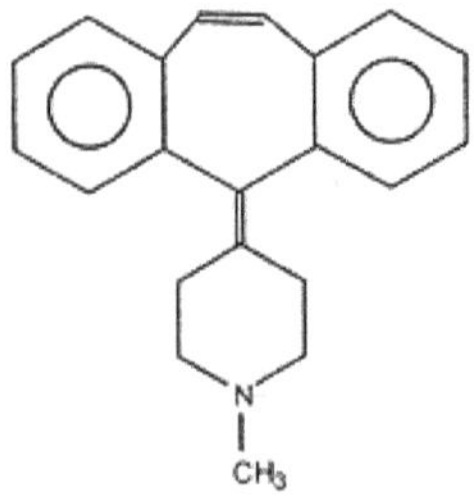

4-(5H) - dibenzo [a,d] cyclohepten - 5- ylidene) -1- methyl piperidine.

Synthesis

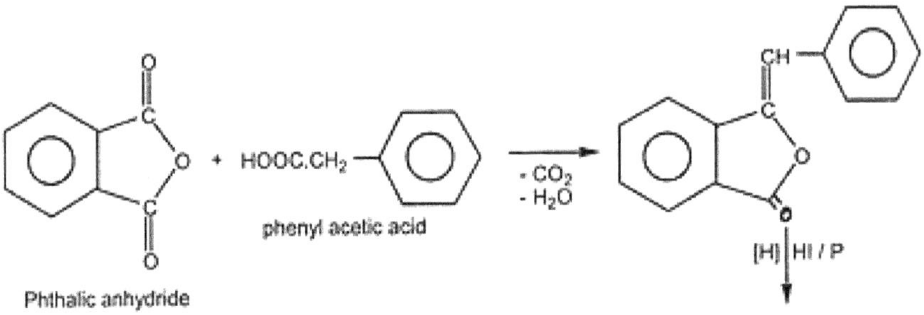

Phthalic anhydride + HOOC.CH$_2$—phenyl acetic acid $\xrightarrow[- H_2O]{- CO_2}$

[H] | HI / P

II. Second Generation H1 Antagonist

Cetirizine hydrochloride (pronounced /s[∪tφrηzi∠n/), an antihistamine, is a major metabolite of hydroxyzine, and a racemic selective H$_1$ receptor inverse agonist used in the treatment of allergies, hay fever, angioedema, and urticaria. The structural similarity of cetirizine to hydroxyzine, and its derivation from piperazine, attribute similar adverse reactions and properties to other piperazine derivatives.

Cetirizine

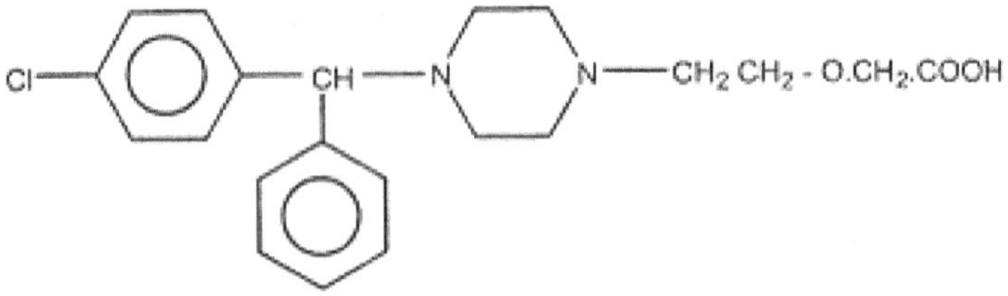

Cl—CH—N—N——CH$_2$ CH$_2$ - O.CH$_2$.COOH

2- {4-[(4- chloro phenyl) phenyl methyl] - 1-piperazinyl ethoxy} acetic acid.

Synthesis

Mechanism of Action for Histamine H1 Antagonist

1. H1 antagonist mainly competitively inhibits the action of histamine on tissues containing H1 receptors.

2. The drugs have the pharmacological actions, opposite to that of histamines and also prevent the access of histamine to its receptors by competitive antagonism.

3. Some antihistamines also antagonize serotonin and bradykinin which are released along with histamine during anaphylaxis reaction.

4. Stimulation of H1 receptors leads to an increase of the intracellular calcium concentration and hydrolysis of phosphatidylino – sitol -4,5-bis phosphate to inositol triphosphate (IP3) and 1,2 diacyl glycerol (DAG). So this histamine induced production of IP3 and DAG is antagonized by H1 receptor antagonist.

5. The increased intra cellular level of IP3 and DAG by histamine is the mobilization of intra cellular calcium.

6. Elevation of intracellular calcium level is associated with various biochemical consequences including the activation of PLC and phospholipase A2, which liberates arachidonic acid from cell membrane, leading to production of powerful mediator like prostacyline and thromboxene A2.

7. In intestine smooth muscles histamine activate the ion channels permeable to Na+ and K+ leading to depolarization and muscle contraction.

8. So the H1 antagonist competitively inhibits the histamine H1 receptor and prevents the above actions.

B Histamine H2 Receptor Antagonist

Ranitidine

Ranitidine hydrochloride is a histamine H_2-receptor antagonist that inhibits stomach acid production. It is commonly used in treatment of peptic ulcer disease (PUD) and gastroesophageal reflux disease (GERD). Ranitidine is also used alongside fexofenadine and other antihistamines for the treatment of skin conditions such as hives. Ranitidine HCl is marketed under the brand name **Zinetac** or **Zantac** (not to be confused with Xanax).

Ranitidine

$CH_2S\ CH_2\ CH_2\ NH\ .C\ .NH.\ CH_3$

$CH.\ NO_2$

CH_3

$N\text{-}CH_2$

CH_3

N-[2-[[[5-(Dimethylamino) methyl]-2 - furanyl]methyl]thio]ethyl]-N'-methyl-2-nitro-1,1 - ethenediamine.

Synthesis

$$\text{furan-CH}_2\text{OH} \xrightarrow[\text{Mannich Reaction}]{(CH_3)_2NH,\ CH_2O} (CH_3)_2N\text{—CH}_2\text{—furan—CH}_2OH \xrightarrow[\text{Cysteamine}]{HSCH_2CH_2NH_2,\ H^+}$$

$$(CH_3)_2N\text{—CH}_2\text{—furan—CH}_2\text{·S—CH}_2CH_2NH_2 \xrightarrow[\substack{CH_3S\text{-}C\text{·}NHCH_3 \\ H\text{-}C\ NO_2}]{} (CH_3)_2N\text{—CH}_2\text{—furan—CH}_2\text{·S—CH}_2CH_2NH\text{-}C\text{—NH-}CH_3$$

$$\overset{\|}{HC}\text{—NO}_2 \quad +\quad CH_3S\text{-}$$

Famotidine

$$CH_2.S.\ CH_2.\ CH_2.NH.C\text{-}NH_2$$
$$\overset{\|}{N}\ SO_2NH_2$$

Thiazole ring with substituents:
NH_2, $C{=}N$, NH_2, S, N

N' - (Amino sulfoxyl) - 3 [[2 - (diamino methylene) - amino] 4 -thiazolyl] methyl] thio] propionamide.

Famotidine is a histamine H_2-receptor antagonist that inhibits stomach acid production, and it is commonly used in the treatment of peptic ulcer disease (PUD) and gastroesophageal reflux disease (GERD/GORD). It is commonly marketed by Johnson & Johnson/Merck under the trade names **Pepcidine** and **Pepcid**. Unlike cimetidine, the first H_2 antagonist, famotidine has no effect on the cytochrome P450 enzyme system, and does not appear to interact with other drugs Certain preparations of famotidine are available over the counter (OTC) in various countries. In the United States, preparations of 10 mg and 20 mg tablets, sometimes in combination with a more traditional antacid, are available OTC. Larger doses still require a prescription. Famotidine is given to surgery patients before operations to prevent post-operation nausea and to reduce the risk of aspiration pneumonitis.

Synthesis

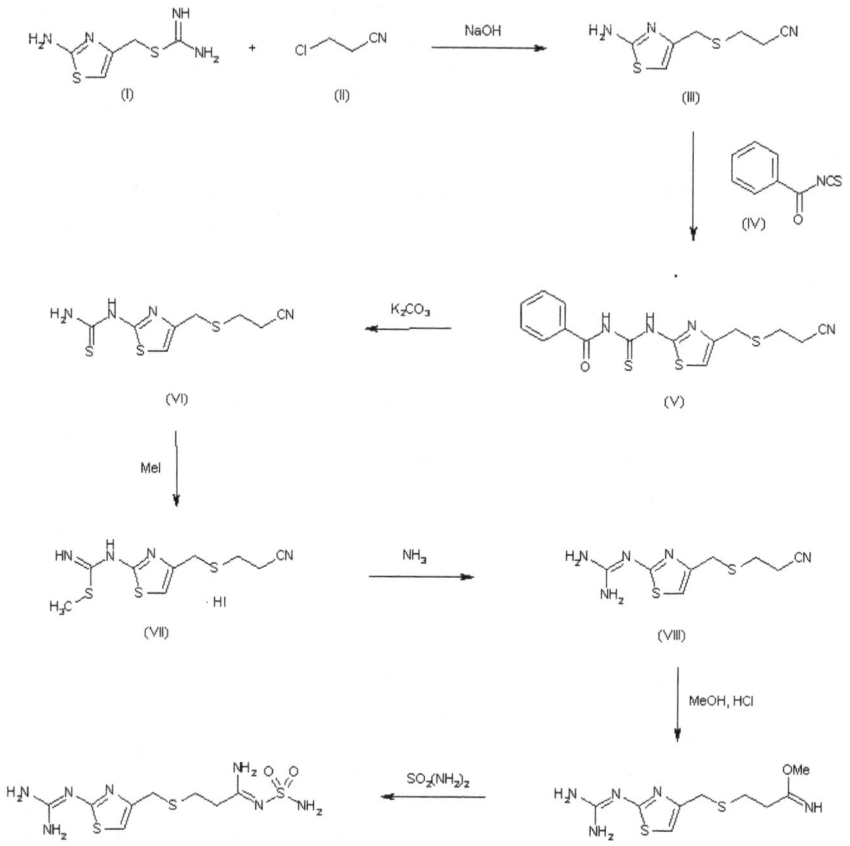

Mechanism of action For Histamine H2 antagonists

1. H2 antagonists mainly antagonize the action of histamine at its H2 receptors which is responsible for acid secretion and peptic ulcer disease.

2. Mucus secreted by the gastric mucous cells combined with surface epithelial bicarbonate secretion contributes to a barrier that prevents gastric acid and pepsin from damaging the gastric mucosa.

3. The acid secretary unit of gastric mucosa is the parietal cell which contains a hydrogen ion pump, H_3O^+ - K^+ - ATP ase system that secretes H3O+ is exchange for the uptake of K^+ ion.

4. Secretion of acid by gastric parietal cells is regulated by the actions of various mediators at receptors including histamine agonism of H2 receptors.

5. Since H2 – receptor antagonists are potent inhibitors of all stimulants of gastric acid secretion, histamine may be considered as a single common final mediator of acid secretion on the parietal cells.

6. The H2 antagonists simple inhibit the direct actions of histamine on acid secretion. H2 antagonists also protect mucosal barriers, proton pump inhibitors, prostaglandins.

SAR for H2 Antagonist

1. Imidazole ring exist in two tautomeric forms. In these form – I to be necessary for maximal H2 antagonist activity.

2. When R is substituted with methyl group, the activity is potentiated.

3. The other heterocyclics like furan, thiazole are enhance the potency and selectivity of H2 receptor antagonism.

4. The ring and terminal nitrogen should be separated by four carbon atoms for optimum activity. The shorter chain decreases the activity.

5. The side chain should contain an electron with drawing substituents and an Isosteric thioether (- S-) link in place of methylene group ($-CH_2$) leads to more active compound.

6. The terminal nitrogen should be polar, non basic substituents for maximal activity.

7. Though a positively charged group binds more tightly to the receptor, it leads to an agonist activity rather than an antagonist activity.

Serotonin

Serotonin is a monoamine neurotransmitter that is primarily found in the gastrointestinal (GI) tract and central nervous system (CNS) of animals. Approximately 80 percent of the human body's total serotonin is located in the enterochromaffin cells in the gut, where it is used to regulate intestinal movements. The remainder is synthesized in serotonergic neurons in the CNS where it has various functions, including the regulation of mood, appetite, sleep, muscle contraction, and some cognitive functions including memory and learning; and in blood platelets where it helps to regulate hemostasis and blood clotting. In addition to animals, serotonin is also found in fungi and plants.

5-hydroxytryptamine, 5-HT, N-methyl-gamma

Biosynthesis

The pathway for the synthesis of serotonin from tryptophan. In animals including humans, serotonin is synthesized from the amino acid L-tryptophan by a short metabolic pathway consisting of two enzymes: tryptophan hydroxylase (TPH) and amino acid decarboxylase (DDC). The TPH-mediated reaction is the rate-limiting step in the pathway. TPH has been shown to exist in two forms: TPH1, found in several tissues, and TPH2, which is a brain-specific isoform. Serotonin taken orally does not pass into the serotonergic pathways of the central nervous system because it does not cross the blood-brain barrier. However, tryptophan and its metabolite 5-hydroxytryptophan (5-HTP), from which serotonin is synthesized, can and do cross the blood-brain barrier. These agents are available as dietary supplements and may be effective serotonergic agents. One product of serotonin breakdown is 5-Hydroxyindoleacetic acid (5 HIAA), which is excreted in the urine.

Tryptophan

Tryptophan hydroxylase

5-hydroxytryptophan

L aromatic acid decarboxylase
(=dopa decarboxylase)

5-hydroxytryptamine (serotonin)

MAO

Monoamine oxidase (MAO)

5-methoxytryptamine

Aldehyde dehydrogenase

5-hydroxyindoleacetic acid (5-HIAA)

Serotonin Antagonist

A **serotonin antagonist** acts to inhibit the action at serotonin receptors. Many of the most important medications of this class selectively act at the 5-HT3 receptor, and thus are known as 5-HT3 antagonists. However, other drugs such as ketanserin act upon other types of serotonin receptors.

Antagonists of the 5-HT_{2A} receptor are sometimes used as atypical antipsychotics, with the typical antipsychotics being purely Dopamine antagonists.

A serotonin receptor agonist **is a compound that activates serotonin receptors, mimicking the effect of the neurotransmitter serotonin. There are various serotonin receptors and ligands.**

5-HT$_{1A}$ receptor

Azapirones such as buspirone, gepirone, and tandospirone are 5-HT$_{1A}$ agonists marketed primarily as anxiolytics, but also recently as antidepressants.

5-HT$_{1B}$ receptor

Triptans such as sumatriptan, rizatriptan, and naratriptan, are 5-HT$_{1B}$ receptor agonists that are used to abort migraine and cluster headache attacks.

5-HT$_{1D}$ receptor

In addition to being 5-HT$_{1B}$ agonists, triptans are also agonists at the 5-HT$_{1D}$ receptor, which contributes to their antimigraine effect.

5-HT$_{1F}$ receptor

LY-334,370 was a selective 5-HT$_{1F}$ agonist that was being developed by Eli Lilly and Company for the treatment of migraine and cluster headaches. Development was halted however due to toxicity detected in animal test subjects.

5-HT$_{2A}$ receptor

Psychedelic drugs such as LSD, mescaline, psilocin, DMT, and 2C-B act as 5-HT$_{2A}$ agonists. Their action at this receptor is responsible for their hallucinogenic effects.

5-HT$_{2C}$ receptor

Lorcaserin is a thermogenic and anorectic weight-loss drug which acts as a selective 5-HT$_{2C}$ agonist.

5-HT$_4$ receptor

Cisapride is a 5-HT$_4$ receptor agonist that has been used to treat disorders of gastrointestinal motility.

5-HT$_7$ receptor

AS-19 (drug) is a 5-HT$_7$ receptor agonist that has been used only in research.

Drugs targeting the 5-HT system

Several classes of drugs target the 5-HT system including some antidepressants, antipsychotics, anxiolytics, antiemetics, and antimigraine drugs as well as the psychedelic drugs and empathogens.

Psychedelic Drugs

The psychedelic drugs psilocin/psilocybin, DMT, mescaline, and LSD are agonists primarily at $5HT_{2A/2C}$ receptors. The Empathogen-entactogen MDMA (ecstasy) releases serotonin from synaptic vesicles of neurons.

Antidepressants

The most prescribed drugs in many parts of the world are drugs which alter serotonin levels. They are used in depression, generalized anxiety disorder and social phobia. The MAOIs prevent the breakdown of monoamine neurotransmitters (including serotonin), and therefore increase concentrations of the neurotransmitter in the brain. MAOI therapy is associated with many adverse drug reactions, and patients are at risk of hypertensive emergency triggered by foods with high tyramine content and certain drugs. Some drugs inhibit the re-uptake of serotonin, making it stay in the synapse longer. The tricyclic antidepressants (TCAs) inhibit the re-uptake of both serotonin and norepinephrine. The newer selective serotonin re-uptake inhibitors (SSRIs) have fewer side-effects and fewer interactions with other drugs. The side effects that have become apparent in recent times include a decrease in bone mass in elderly and increased risk for osteoporosis. However, it is not yet clear whether it is due to SSRI action on peripheral serotonin production and or action in the gut or in the brain. Certain SSRI medications have been shown to lower serotonin levels below the baseline after chronic use, despite initial increases in serotonin. This has been connected to the observation that the benefit of SSRI's may decrease in selected patients after a long-term treatment. A switch in medication will usually resolve this issue (up to 70% of the time). The novel antidepressant tianeptine, a selective serotonin reuptake *enhancer*, has mood-elevating effects. This provides evidence for the theory that serotonin is most likely used to regulate the extent or intensity of moods.

Althrough phobias and depression might be attenuated by serotonin-altering-drugs this does not mean that the individuals' situation has been improved, but only the individuals perception of the environment. Sometimes a lower serotonin level might be beneficial, for example in the ultimatum game, where players with normal serotonin levels are more prone to accept unfair offers than participants whose serotonin levels have been artificially lowered.

Serotonin syndrome

Extremely high levels of serotonin can have toxic and potentially fatal effects, causing a condition known as serotonin syndrome. In practice, such toxic levels are essentially impossible to reach through an overdose of a single anti-depressant drug, but require a combination of serotonergic agents, such as an SSRI with an MAOI. The intensity of the symptoms of serotonin syndrome varies over a wide spectrum, and the milder forms are seen even at non-toxic levels.

Antiemetics

5-HT_3 antagonists such as ondansetron, granisetron, and tropisetron are important antiemetic agents. They are particularly important in treating the nausea and vomiting that occur during anticancer chemotherapy using cytotoxic drugs. Another application is in the treatment of post-operative nausea and vomiting.

Angiotensin

Angiotensin, a protein, causes blood vessels to constrict, and drives blood pressure up. It is part of the renin-angiotensin system, which is a major target for drugs that lower blood pressure. Angiotensin also stimulates the release of aldosterone from the adrenal cortex.

Aldosterone promotes sodium retention in the distal nephron, in the kidney, which also drives blood pressure up.

Angiotensin is an oligopeptide in the blood that causes vasoconstriction, increased blood pressure, and release of aldosterone from the adrenal cortex. It is a hormone and a powerful dipsogen. It is derived from the precursor molecule angiotensinogen, a serum globulin produced in the liver. It plays an important role in the renin-angiotensin system.

Angiotensin was independently isolated in Indianapolis and Argentina in the late 1930s (as 'Angiotonin' and 'Hypertensin' respectively) and subsequently characterised and synthesized by groups at the Cleveland Clinic and Ciba laboratories in Basel, Switzerland.

Renin-angiotensin-aldosterone system

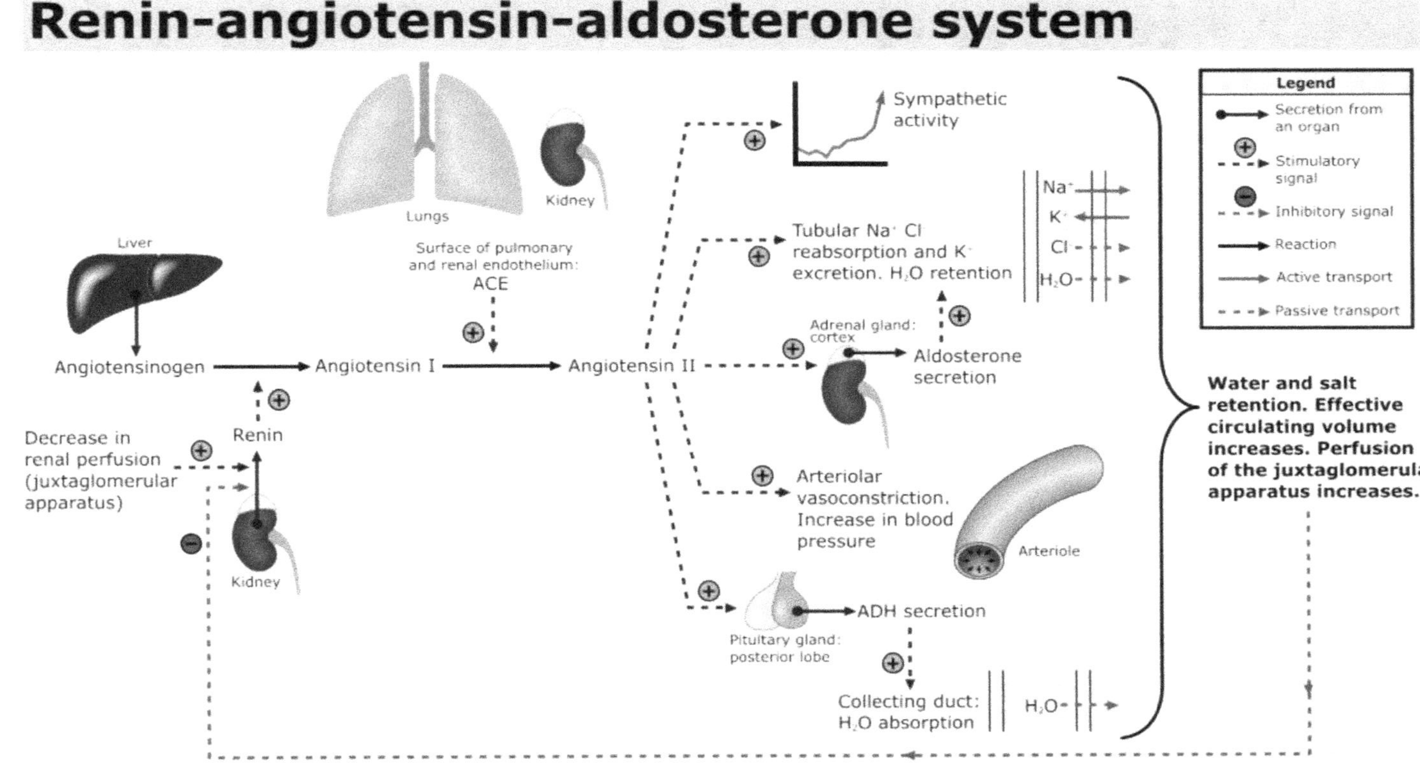

Angiotensin II receptor antagonist

Angiotensin II receptor antagonists, also known as **angiotensin receptor blockers (ARBs)**, **AT_1-receptor antagonists** or **sartans**, are a group of pharmaceuticals which modulate the renin-angiotensin-aldosterone system. Their main use is in hypertension (high blood pressure), diabetic nephropathy (kidney damage due to diabetes) and congestive heart failure.

Losartan, first Drug of this class

Mechanism of action

These substances are AT_1-receptor antagonists – that is, they block the activation of angiotensin II AT_1 receptors. Blockade of AT_1 receptors directly causes vasodilation, reduces secretion of vasopressin, reduces production and secretion of aldosterone, amongst other actions – the combined effect of which is reduction of blood pressure.

The specific efficacy of each ARB within this class is made up of a combination of three pharmacodynamic and pharmacokinetic parameters. For these three key PD/ PK areas that indicate efficacy, it is important to see that one needs a combination of all three at an effective level; the parameters of the three characteristics will need to compiled into a table similar too one below, eliminating duplications and arriving at consensus values; the latter are at variance now.

Uses

Angiotensin II receptor antagonists are primarily used for the treatment of hypertension where the patient is intolerant of ACE inhibitor therapy. They do not inhibit the breakdown of bradykinin or other kinins, and are thus only rarely associated with the persistent dry cough and/or angioedema that limit ACE inhibitor therapy. More recently, they have been used for the treatment of heart failure in patients intolerant of ACE inhibitor therapy, particularly candesartan. Irbesartan and losartan have trial data showing benefit in hypertensive patients with type II diabetes, and may delay the progression of diabetic nephropathy. Candesartan is used experimentally in preventive treatment of migraine. Another angiotensin II receptor antagonist, olmesartan, is an important part of the Marshall Protocol, invented by Australian biomedical researcher Trevor Marshall.

The angiotensin II receptor blockers have differing potencies in relation to blood pressure control, with statistically differing blood pressure effects at the maximal doses when used in clinical practice, the particular agent used may vary based on the degree of blood pressure response required.

Prostaglandin

Prostaglandin is any member of a group of lipid compounds that are derived enzymatically from fatty acids and have important functions in the animal body. Every prostaglandin contains 20 carbon atoms, including a 5-carbon ring. They are mediators and have a variety of strong physiological effects, such as regulating the contraction and relaxation of smooth muscle tissue. Although they are technically hormones, they are rarely classified as such.The prostaglandins, together with the thromboxanes and prostacyclins, form the prostanoid class of fatty acid derivatives. The prostanoid is a subclass of eicosanoids.

Function

There are currently ten known prostaglandin receptors on various cell types. Prostaglandins ligate a sub-family of cell surface seven-transmembrane receptors,   G-protein-coupled receptors. These receptors are termed DP1-2, EP1-4, FP, IP1-2, and TP, corresponding to the receptor that ligates the corresponding prostaglandin (e.g., DP1-2 receptors bind to PGD2).

The diversity of receptors means that prostaglandins act on an array of cells and have a wide variety of effects:

- cause constriction or dilation in vascular smooth muscle cells
- cause aggregation or disaggregation of platelets
- sensitize spinal neurons to pain
- decrease intraocular pressure
- regulate inflammatory mediation

- regulate calcium movement
- Control hormone regulation
- Control cell growth

Prostaglandins are potent but have a short half-life before being inactivated and excreted. Therefore, they send only paracrine (locally active) or autocrine (acting on the same cell from which it is synthesized) signals

Types

Different types of prostaglandin, prostaglandin I_2 (PGI$_2$), prostaglandin E_2 (PGE$_2$), and prostaglandin $F_{2\alpha}$ (PGF$_{2\alpha}$).

Role in pharmacology

Inhibition

Examples of prostaglandin antagonists are:

- NSAIDs (inhibit cyclooxygenase)
- Corticosteroids (inhibit phospholipase A2 production)
- COX-2 selective inhibitors or coxibs
- Cyclopentenone prostaglandins may play a role in inhibiting inflammation

Clinical uses

Synthetic prostaglandins are used:

- To induce childbirth (parturition) or abortion (PGE$_2$ or PGF$_2$, with or without mifepristone, a progesterone antagonist);
- To prevent closure of patent ductus arteriosus in newborns with particular cyanotic heart defects (PGE$_1$)
- To prevent and treat peptic ulcers (PGE)
- As a vasodilator in severe Raynaud's phenomenon or ischemia of a limb
- In pulmonary hypertension
- In treatment of glaucoma (as in bimatoprost ophthalmic solution, a synthetic prostamide analog with ocular hypotensive activity)
- To treat erectile dysfunction or in penile rehabilitation following surgery (PGE1 as alprostadil).
- To treat egg binding in small birds
- As an ingredient in eyelash and eyebrow growth beauty products due to side effects associated with increased hair growth

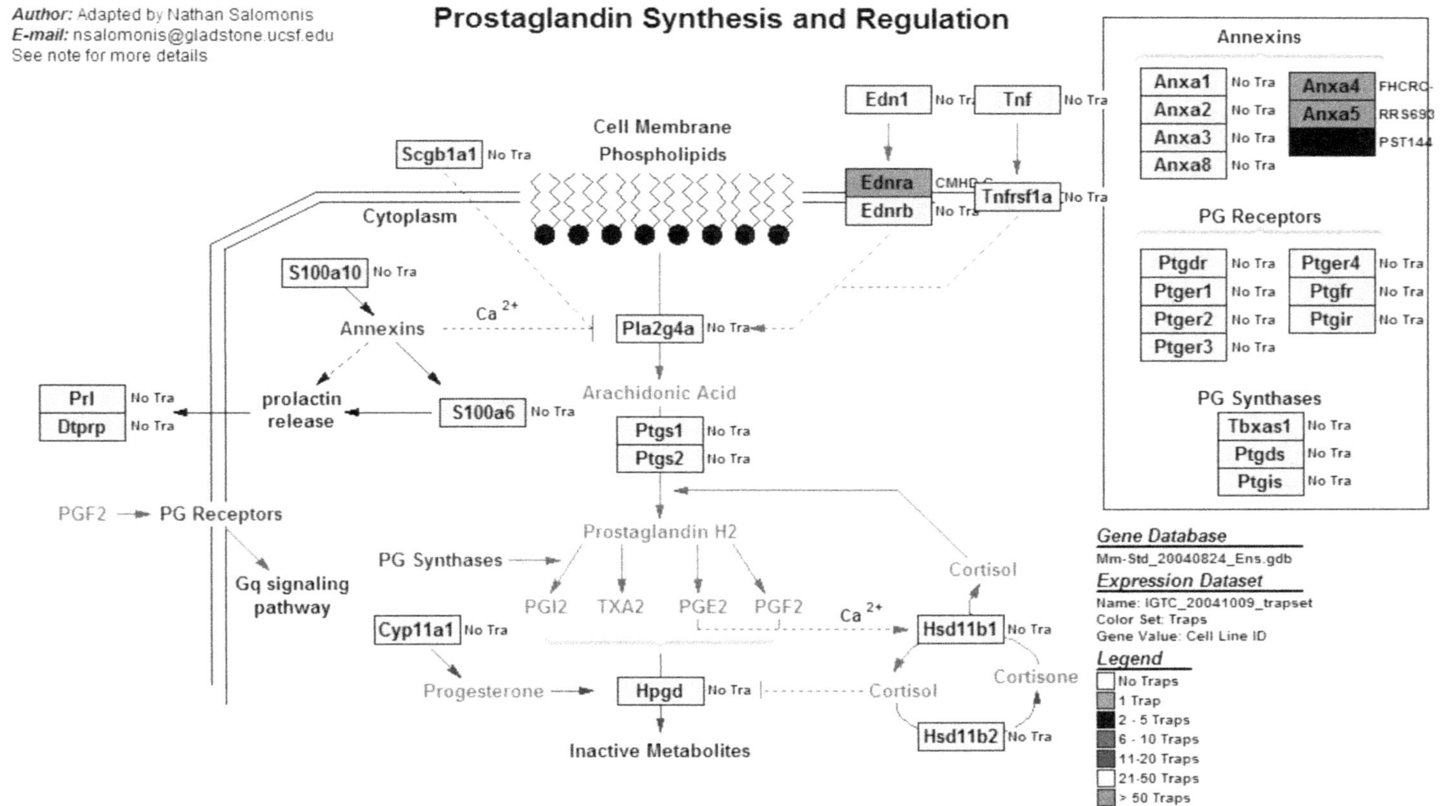
Author: Adapted by Nathan Salomonis
E-mail: nsalomonis@gladstone.ucsf.edu
See note for more details
Prostaglandin Synthesis and Regulation
Cell Membrane Phospholipids
Scgb1a1 No Tra
Cytoplasm
S100a10 No Tra
Ca 2+
Annexins
Prl No Tra
Dtprp No Tra
prolactin release
S100a6 No Tra
Pla2g4a No Tra
Arachidonic Acid
Ptgs1 No Tra
Ptgs2 No Tra
PGF2 PG Receptors
Gq signaling pathway
Prostaglandin H2
PG Synthases
PGI2 TXA2 PGE2 PGF2
Ca 2+
Cyp11a1 No Tra
Progesterone
Hpgd No Tra
Cortisol
Cortisone
Hsd11b1 No Tra
Hsd11b2 No Tra
Inactive Metabolites
Edn1 No Tra
Tnf No Tra
Ednra
Ednrb No Tra
CMHD-C
Tnfrsf1a No Tra
Annexins
Anxa1 No Tra
Anxa2 No Tra
Anxa3 No Tra
Anxa8 No Tra
Anxa4 FHCRC-
Anxa5 RRS693
PST144
PG Receptors
Ptgdr No Tra
Ptger1 No Tra
Ptger2 No Tra
Ptger3 No Tra
Ptger4 No Tra
Ptgfr No Tra
Ptgir No Tra
PG Synthases
Tbxas1 No Tra
Ptgds No Tra
Ptgis No Tra
Gene Database
Mm-Std_20040824_Ens.gdb
Expression Dataset
Name: IGTC_20041009_trapset
Color Set: Traps
Gene Value: Cell Line ID
Legend
No Traps
1 Trap
2 - 5 Traps
6 - 10 Traps
11-20 Traps
21-50 Traps
> 50 Traps

MesoProstal

Misoprostol is a drug that is used for the prevention of non-steroidal anti-inflammatory drug (NSAID)-induced gastric ulcers, for early abortion, to treat missed miscarriage, and to induce labor. The last use is controversial in the United States. Misoprostol was invented and marketed by G.D. Searle & Company (now Pfizer) under the trade name **Cytotec** (often misspelled **Cyotec**), but other brand-name and generic formulations are now available as well. Chemically, misoprostol is a synthetic prostaglandin E_1 (PGE_1) analogue.

Methyl7-(((1R,2R,3R)-3-hydroxy-2-((S,E)-4-hydroxy-4-methyloct-1-enyl)-5-oxocyclopentyl) heptanoate

Ulcer Prevention

Misoprostol is approved for use in the prevention of NSAID-induced gastric ulcers. It acts upon gastric parietal cells, inhibiting the secretion of gastric acid via G-protein coupled receptor-mediated inhibition of adenylate cyclase, which leads to decreased intracellular cyclic AMP levels and decreased proton pump activity at the apical surface of the parietal cell. Because other classes of drugs, especially H2-receptor antagonists and proton pump inhibitors, are more effective for the treatment of acute peptic ulcers, Misoprostol is only indicated for use by people who are both taking NSAIDs and are at high risk for NSAID-induced ulcers, including the elderly and people with ulcer complications. Misoprostol is sometimes co-prescribed with NSAIDs to prevent their common adverse effect of gastric ulceration (e.g. with Diclofenac in Arthrotec). Misoprostol has other protective actions, but is only clinically effective at doses high enough to reduce gastric acid secretion. For instance, at lower doses misoprostol may stimulate increased secretion of the protective mucus that lines the gastrointestinal tract and increase mucosal blood flow, thereby increasing mucosal integrity—however, these effects are not pronounced enough to warrant prescription of misoprostol at doses lower than those needed to achieve gastric acid suppression.

Labor Induction

Misoprostol is commonly used for labor induction. It causes uterine contractions and the ripening (effacement or thinning) of the cervix. Misoprostol is more effective in starting labor than other drugs used for labor induction. It is also significantly less expensive than the other commonly used ripening agent, dinoprostone (trade names *Cervidil* and *Prepidil*).

Oxytocin (trade names *Pitocin* and *Syntocinon*) has long been used as the standard agent for labor induction, but doesn't work well when the cervix is not yet ripe. In addition to being used alone to induce labor, misoprostol may be used in conjunction with oxytocin. Protocols for inducing labor with misoprostol typically call for 25 µg to be administered vaginally. In countries where the only approved use of misoprostol is ulcer prevention, misoprostol is not sold in tablets smaller than 100 µg. When used for induction, the 100 µg tablet is commonly split into two or four pieces

Induced Abortion

Earlier pregnancy

Misoprostol is one of the drugs used for medical abortions in lieu of surgical evacuation. The advantages of medical abortion over surgical abortion include reduced invasiveness of the procedure, lack of risks from general anesthesia (which is often used for surgical abortions), and lack of risk of secondary infertility due to scarring and intrauterine adhesions (Asherman's Syndrome). Furthermore, it is less complicated to administer and less expensive.

In many countries' it is used in conjunction with mifepristone (RU-486). After mifepristone is taken orally, misoprostol is taken 24–72 hours later, causing the expulsion of the embryo and associated matter in approximately 92% of the cases. No large studies have established a protocol for the use of misoprostol alone, and the range of efficacy is 65%–93% depending on sample size, gestational age, and other test variables; Misoprostol alone may be more effective in earlier gestation. The side effects associated with the misoprostol-only regimen are generally much more severe than those associated with the combined regimens. Misoprostol is used for self-induced abortions in Brazil, where black market prices exceed US $100 per dose. Illegal medically-unsupervised misoprostol abortions in Brazil are associated with a lower complication rate than other forms of illegal self-induced abortion, but are still associated with a higher complication rate than legal, medically supervised surgical and chemical abortions. Failed misoprostol abortions are associated with birth defects in some cases Poor immigrant populations in New York have also been observed to use self-administered misoprostol to induce abortions, as this method is much cheaper than a surgical abortion (about $2 per dose).

Later pregnancy

Misoprostol can also be used to dilate the cervix in preparation for a surgical abortion, particularly in the second trimester (either alone or in combination with laminaria stents).

Missed miscarriage

Misoprostol is sometimes used to treat early fetal death in the absence of spontaneous miscarriage, but further research is needed to establish a safe, effective protocol.

Post-partum hemorrhage

Misoprostol is also used to prevent and treat post-partum hemorrhage, but it has more side effects and is less effective than oxytocin for this purpose. However, it is inexpensive and thermostable (thus does not require refrigeration like oxytocin) making it a cost-effective and valuable drug to use in the developing world.

Other Gynecological uses

Although the practice remains uncommon, some gynecologists are now using low doses of misoprostol to soften the cervix prior to the insertion of intrauterine devices (especially in nulliparous women where insertion may be challenging).

Erectile dysfunction

A 1998 study found misoprostol to be helpful as a supplement to a vacuum pump (VED) in the treatment of erectile dysfunction, but not effective by itself. The paper concluded "The intraurethral application of misoprostol significantly improves the quality of VED-induced erections. This agent seems to be a cheap intraurethral adjunct to VED with mild to moderate local side-effects".

Carboprost

Carboprost is a synthetic prostaglandin analogue of $PGF_{2\alpha}$ (specifically, it is 15-methyl-$PGF_{2\alpha}$) with oxytocic properties. The trade name of carboprost tromethamine is **Hemabate**. Carboprost induces contractions and can trigger abortion in early pregnancy. It also reduces postpartum bleeding. It is used for the treatment of molar pregnancy. Exert caution in asthmatic patients.

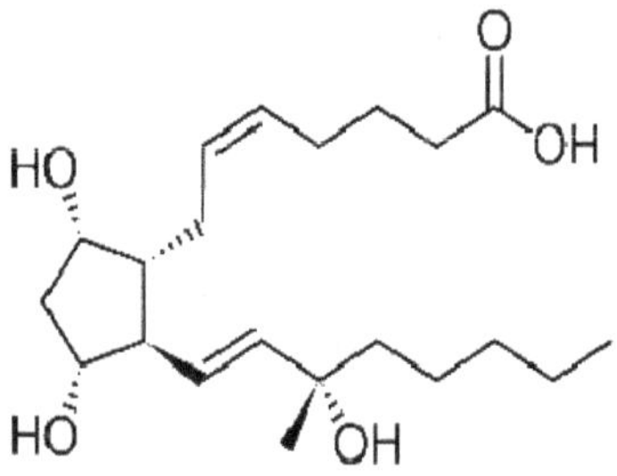

(5Z,9α,11α,13E,15S)-9,11,15-trihydroxy-15- methylprosta-5,13-dien-1-oic acid

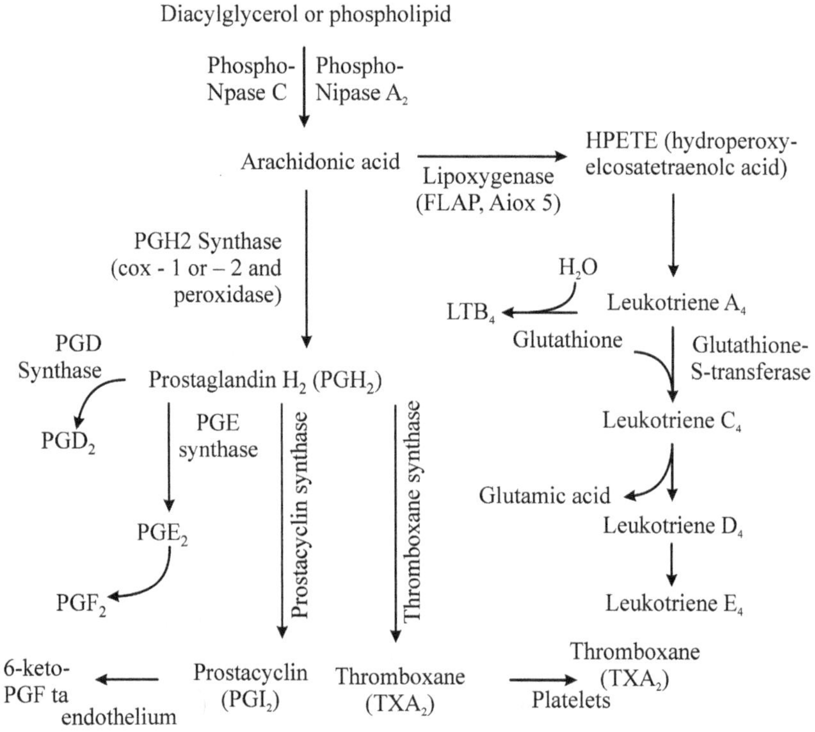

Diacylglycerol or phospholipid

Phospho-Npase C | Phospho-Nipase A$_2$

Arachidonic acid → Lipoxygenase (FLAP, Aiox 5) → HPETE (hydroperoxy-elcosatetraenolc acid)

PGH2 Synthase (cox - 1 or – 2 and peroxidase)

H_2O

LTB$_4$ ← Leukotriene A$_4$

Glutathione

Glutathione-S-transferase

PGD Synthase

Prostaglandin H$_2$ (PGH$_2$)

PGD$_2$

PGE synthase

Prostacyclin synthase

Thromboxane synthase

Leukotriene C$_4$

Glutamic acid ← Leukotriene D$_4$

PGE$_2$

Leukotriene E$_4$

PGF$_2$

Thromboxane (TXA$_2$)

6-keto-PGF ta endothelium ← Prostacyclin (PGI$_2$)   Thromboxane (TXA$_2$) → Platelets